Fifty Years of Free Radicals

Fifty Years of Free Radicals

Cheves Walling

PROFILES, PATHWAYS, AND DREAMS
Autobiographies of Eminent Chemists

Jeffrey I. Seeman, Series Editor

American Chemical Society, Washington, DC 1995

Library of Congress Cataloging-in-Publication Data

Walling, Cheves, 1916–
Fifty years of free radicals / Cheves Walling.

p. cm. — (Profiles, pathways, and dreams, ISSN 1047–8329)
Includes bibliographical references (p. –) and index.
ISBN 0–8412–1830–7

1. Walling, Cheves, 1916– . 2. Chemists—United States—Biography. 3. Free radicals (Chemistry)

I. Title. II. Series.

QD22.W27A3 1995
540'.92—
dc20 95–20101
[B] CIP
Jeffrey I. Seeman, Series Editor

This book is printed on acid-free, recycled paper.

PRINTED IN THE UNITED STATES OF AMERICA

1995 Advisory Board

Foreword

In 1986, the ACS Books Department accepted for publication a collection of autobiographies of organic chemists, to be published in a single volume. However, the authors were much more prolific than the project's editor, Jeffrey I. Seeman, had anticipated, and under his guidance and encouragement, the project took on a life of its own. The original volume evolved into 22 volumes, and the first volume of Profiles, Pathways, and Dreams: Autobiographies of Eminent Chemists was published in 1990. Unlike the original volume, the series was structured to include chemical scientists in all specialties, not just organic chemistry. Our hope is that those who know the authors will be confirmed in their admiration for them, and that those who do not know them will find these eminent scientists a source of inspiration and encouragement, not only in any scientific endeavors, but also in life.

Contributors

We thank the following corporations and Herchel Smith for their generous financial support of the series Profiles, Pathways, and Dreams.

Akzo nv

Bachem Inc.

DuPont

Duphar B.V.

Eisai Co., Ltd.

Fujisawa Pharmaceutical Co., Ltd.

Hoechst Celanese Corporation

Imperial Chemical Industries PLC

Kao Corporation

Mitsui Petrochemical Industries, Ltd.

The NutraSweet Company

Organon International B.V.

Pergamon Press PLC

Pfizer Inc.

Philip Morris

Quest International

Sandoz Pharmaceuticals Corporation

Sankyo Company, Ltd.

Schering–Plough Corporation

Shionogi Research Laboratories, Shionogi & Co., Ltd.

Herchel Smith

Suntory Institute for Bioorganic Research

Takasago International Corporation

Takeda Chemical Industries, Ltd.

Unilever Research U.S., Inc.

Profiles, Pathways, and Dreams

Titles in This Series

Sir Derek H. R. Barton *Some Recollections of Gap Jumping*

Arthur J. Birch *To See the Obvious*

Melvin Calvin *Following the Trail of Light: A Scientific Odyssey*

Donald J. Cram *From Design to Discovery*

Michael J. S. Dewar *A Semiempirical Life*

Carl Djerassi *Steroids Made It Possible*

Ernest L. Eliel *From Cologne to Chapel Hill*

Egbert Havinga *Enjoying Organic Chemistry, 1927–1987*

Rolf Huisgen *The Adventure Playground of Mechanisms and Novel Reactions*

William S. Johnson *A Fifty-Year Love Affair with Organic Chemistry*

Raymond U. Lemieux *Explorations with Sugars: How Sweet It Was*

Herman Mark *From Small Organic Molecules to Large: A Century of Progress*

Bruce Merrifield *Life During a Golden Age of Peptide Chemistry: The Concept and Development of Solid-Phase Peptide Synthesis*

Teruaki Mukaiyama *To Catch the Interesting While Running*

Koji Nakanishi *A Wandering Natural Products Chemist*

Tetsuo Nozoe *Seventy Years in Organic Chemistry*

Vladimir Prelog *My 132 Semesters of Chemistry Studies*

John D. Roberts *The Right Place at the Right Time*

Paul von Rague Schleyer *From the Ivy League into the Honey Pot*

F. Gordon A. Stone *Leaving No Stone Unturned: Pathways in Organometallic Chemistry*

Andrew Streitwieser, Jr. *A Lifetime of Synergy with Theory and Experiment*

Cheves Walling *Fifty Years of Free Radicals*

About the Editor

JEFFREY I. SEEMAN received his B.S. with high honors in 1967 from the Stevens Institute of Technology in Hoboken, New Jersey, and his Ph.D. in organic chemistry in 1971 from the University of California, Berkeley. Following a two-year staff fellowship at the Laboratory of Chemical Physics of the National Institutes of Health in Bethesda, Maryland, he joined the Philip Morris Research Center in Richmond, Virginia. In 1983–1984, he enjoyed a sabbatical year at the Dyson Perrins Laboratory in Oxford, England, and claims to have visited more than 90% of the castles in England, Wales, and Scotland.

Seeman's 90 published papers include research and patents in the areas of photochemistry, nicotine and tobacco alkaloid chemistry and synthesis, conformational analysis, pyrolysis chemistry, organotransition metal chemistry, the use of cyclodextrins for chiral recognition, and structure–activity relationships in olfaction. He was a plenary lecturer at the Eighth IUPAC Conference on Physical Organic Chemistry held in Tokyo in 1986 and has been an invited lecturer at numerous scientific meetings and universities. From 1989–1994 Seeman served on the Petroleum Research Fund Advisory Board. He continues to count Nero Wolfe and Archie Goodwin among his best friends.

Contents

Photographs xiii

Preface xvii

Editor's Note xxi

Early Life and Education 1
  My Introduction to Chemistry 1

Graduate Work at Chicago, 1937–1939 13
  Kharasch's Research Group 15
  Important Influences 16

Starting Out in Industry 19
  E. I. du Pont de Nemours and Company, 1939–1943 19
  Polymer Chemistry: The U.S. Rubber Company, 1943–1949 22
  Structure and Reactivity in Radical Chemistry 27
  Other Developments in Free Radical Chemistry 32
  Free Radical Additions 35
  Autoxidation: The Faraday Society Discussion, 1947 36
  Lever Brothers Company, 1949–1952 40

A Goal Realized—Academic Research Columbia University, 1952–1969 45

High-Pressure Reactions 45
Free Radicals in Solution 49
New Lines of Free Radical Research 50
Hypochlorite Chemistry and Halogen Carriers 53
Thiyl and Phosphoranyl Radicals 57
Radical Cyclizations 58
Peroxide Chemistry 61
The Academic Scene at Columbia 66

Branching Out 69
Committee on Professional Training 69
Research Groups 71
Other Involvements 73
Travel 77

The University of Utah, 1969 to the Present 82
Chemically Induced Dynamic Nuclear Polarization 86
Hydroxyl Radical Chemistry 87
Radical Cations 90
My Adventures as an Editor 93
Other Adventures 98
"Cold Fusion" 103
An Overview of Free Radical Chemistry 114
Ethics in Science 117
A Final Word 119

References 123

Index 131

Photographs

The author at age five in an uncharacteristic pose 2

Senior year in high school, spring 1933 5

At Harvard, 1936 7

At French Lick, Indiana, 1937 9

At the 2nd Reaction Mechanisms Conference at Colby Junior College, New London, New Hampshire, 1948 17

Walling family in Montclair, New Jersey, 1946 21

Farewell luncheon at U.S. Rubber, 1949 25

At Gordon Research Conference, New Hampton, New Hampshire, 1971 27

Frank Mayo and Paul Bartlett at the Stanford Research Institute in 1968 31

With Frank Mayo at U.S. Rubber in 1947 37

Oxford University, England, 1960 39

Participants at the 3rd Reaction Mechanisms Conference, Northwestern University, 1950 42

Christopher Ingold at 3rd Reaction Mechanisms Conference, Northwestern University, 1950 43

Louis Plack Hammett, October 1954 44

At the high-pressure laboratory at Columbia University, 1955 46

With M. G. Gonikberg in Moscow, 1960. 48

Glen Russell, Peter Wagner, and Bill Pryor at the ACS spring meeting in St. Louis, 1984. 52

Former research associates at a symposium at Park City, Utah, 1986. 54

Marc Julia, Mme. Julia, and Jane Walling in Park City, Utah, 1980. 59

Chengxue Zhao and Cheves Walling at the University of Utah 61

Charlie Dawson, Nick Turro, and Ronald Breslow, 1968 67

With John Bailar in 1972 69

Jane and Barbara Walling with Ed and Mrs. Wiig in Fairbanks, Alaska, 1967 70

Changing a tire in the Nevada desert, 1958 77

Generals George Marshall and Omar Bradley at Harvard Commencement in 1947 78

Leopold Ruzicka and Vladimir Prelog at E.T.H. in Zurich, 1960 78

At the National Science Foundation Summer Institute in Durango, Colorado, 1960 79

With Ed King at Hallett Peak, Colorado, 1965 80

With Sadie Walton and Jane Walling in Mexico City, 1975 80

Fred Hawthorne, Riverside, California, 1967 81

With Bob Parry and Henry Eyring, University of Utah, 1970 83

Jean-Marie Lehn at Arches National Park, Utah, 1980 85

With son Cheves Walling at Mt. Ellen, Utah, 1972 86

Ed and Dianne Schulman, Salt Lake City, Utah, 1973 87

With Fred Greene in 1978 94

With Charlotte Sauer, University of Utah, 1989 95

At a nightclub near Cairo, Egypt, 1974 98

With Jane Walling in Japan, 1978 99

With W. Ando in Tsukuba, Japan, 1978 99

With Hideki Sakurai in Salt Lake City, Utah, 1979 100

At Lanzhou University, People's Republic of China, 1982 100

With Chengxue Zhao at the Great Wall of China, 1982 101

At a luncheon in Washington, DC, 1983. 101

A birthday greeting from Athel Beckwith's research group at the Australian National University in Canberra, Australia, 1987 103

Tito Scaiano and Wes Bentrude, Zurich, 1988. 104

With Ellie and Frank Mayo at Yosemite National Park, California, 1979. 105

At the International Symposium on Free Radicals, Freiburg, Germany, 1981. 106

At the 3rd International Symposium on Organic Free Radicals in St. Andrews, Scotland, 1984. 106

At Walling's 70th birthday symposium, Park City, Utah, 1986. 107

With family members at my 70th birthday symposium, Park City, Utah, 1986. 107

With Jane Walling in Indonesia, 1989. 110

With Jane Walling in Jaffrey, New Hampshire, 1991. 120

Preface

"HOW DID YOU GET THE IDEA—and the good fortune—to convince 22 world-famous chemists to write their autobiographies?" This question has been asked of me, in these or similar words, frequently over the past several years. I hope to explain in this preface how the project came about, how the contributors were chosen, what the editorial ground rules were, what was the editorial context in which these scientists wrote their stories, and the answers to related issues. Furthermore, several authors specifically requested that the project's boundary conditions be known.

As I was preparing an article[1] for *Chemical Reviews* on the Curtin–Hammett principle, I became interested in the people who did the work and the human side of the scientific developments. I am a chemist, and I also have a deep appreciation of history, especially in the sense of individual accomplishments. Readers' responses to the historical section of that review encouraged me to take an active interest in the history of chemistry. The concept for Profiles, Pathways, and Dreams resulted from that interest.

My goal for Profiles was to document the development of modern organic chemistry by having individual chemists discuss their roles in this development. Authors were not chosen to represent my choice of the world's "best" organic chemists, as one might choose the "baseball all-star team of the century". Such an attempt would be foolish: Even the selection committees for the Nobel prizes do not make their decisions on such a premise.

The selection criteria were numerous. Each individual had to have made seminal contributions to organic chemistry over a multidecade career. (The average age of the authors is over 70!) Profiles would represent scientists born and professionally productive in different countries. (Chemistry in 13 countries is detailed.) Taken together, these individuals were to have conducted research in nearly all subspecialties of organic chemistry. Invitations to contribute were based on solicited advice and on recommendations of chemists from five continents, including nearly all of the contributors. The final assemblage was selected entirely and exclusively by me. Not all who were invited chose to participate, and not all who should have been invited could be asked.

A very detailed four-page document was sent to the contributors, in which they were informed that the objectives of the series were

1. to delineate the overall scientific development of organic chemistry during the past 30–40 years, a period during which this field has dramatically changed and matured;
2. to describe the development of specific areas of organic chemistry; to highlight the crucial discoveries and to examine the impact they have had on the continuing development in the field;
3. to focus attention on the research of some of the seminal contributors to organic chemistry; to indicate how their research programs progressed over a 20–40-year period; and
4. to provide a documented source for individuals interested in the hows and whys of the development of modern organic chemistry.

One noted scientist explained his refusal to contribute a volume by saying, in part, that "it is extraordinarily difficult to write in good taste about oneself. Only if one can manage a humorous and light touch does it come off well. Naturally, I would like to place my work in what I consider its true scientific perspective, but . . ."

Each autobiography reflects the author's science, his lifestyle, and the style of his research. Naturally, the volumes are not uniform, although each author attempted to follow the guidelines. "To write in good taste" was not an objective of the series. On the contrary, the authors were specifically requested not to write a review article of their field, but to detail their own research accomplishments. To the extent that this instruction was followed and the result is not "in good taste", then these are criticisms that I, as editor, must bear, not the writer.

As in any project, I have a few regrets. It is truly sad that Egbert Havinga and Herman Mark, who each wrote a volume, and David Ginsburg, who translated another, died during the course of this project. There have been many rewards, some of which are documented in my personal account of this project, entitled "Extracting the Essence: Adventures of an Editor" published in *CHEMTECH*.[2]

Acknowledgments

I join the entire scientific community in offering each author unbounded thanks. I thank their families and their secretaries for their contributions. Furthermore, I thank numerous chemists for reading and reviewing the autobiographies, for lending photographs, for sharing information, and for providing each of the authors and me the encouragement to proceed in a project that was far more costly in time and energy than any of us had anticipated.

I thank my employer, Philip Morris USA, and J. Charles, R. N. Ferguson, K. Houghton, H. Grubbs, and W. F. Kuhn, for without their support Profiles, Pathways, and Dreams could not have been. I thank ACS Books, and in particular, Robin Giroux (production manager), Janet Dodd (senior editor), Joan Comstock (department head), and their staff for their hard work, dedication, and support. Each reader no doubt joins me in thanking 24 corporations and Herchel Smith for financial support for the project.

I thank my children, Jonathan and Brooke, for their patience and understanding; remarkably, I have been working on Profiles for more than half of their lives—probably the only half that they can remember! Finally, I again thank all those mentioned and especially my family, friends, colleagues, and the 22 authors for allowing me to share this experience with them.

JEFFREY I. SEEMAN
Philip Morris Research Center
Richmond, VA 23234

April 19, 1993

[1] Seeman, J. I. *Chem. Rev.* **1983**, *83*, 83–134.
[2] Seeman, J. I. *CHEMTECH* **1990**, *20*(2), 86–90.

Editor's Note

"CHEVES WALLING IS ONE OF THE FOUNDING FATHERS OF RADICAL CHEMISTRY," declared Albert Padwa, a professor at Emory and former student from Columbia. "He helped set the parameters twenty years before the explosive growth of the field. His work pervaded the thinking in radical chemistry." The current trend toward using free radicals in synthetic organic chemistry is based on Walling's pioneering work. "It could have been done 25 years earlier if they had talked with Cheves," joked Padwa. "Yes, radicals are not in general very selective; they are unusually reactive; but they are quite valuable in the craft that is synthesis. It just took many years for synthetic chemistry to catch up with Walling's contributions."

Undoubtedly Walling's time in industry helped connect the fundamental science with commercial practicality. After graduate school, he began his professional career with DuPont (1939–1943), then moved to the U.S. Rubber Company (1943–1949) and then Lever Brothers (1949–1952) before realizing his long-term goal of university-based basic research. It was in 1952, Walling recalls, that "Louis Hammett [Chairman of the Department of Chemistry at Columbia University in New York] and I had discussed academic positions . . . and an offer was made."

"Louis Hammett was my model. He was quiet and rather aloof, very honest and very straightforward. He was an admired elder, and I got to observe him closely when he was chairman of the department. Three other people whom I admire very much and who had a lasting influence on me were Frank Mayo, Paul Bartlett and Frank Westheimer. Their approach to science . . . their honest, thorough, and careful science, was an inspiration."

Walling, as professor and scientist, worked in an atmosphere very different from today. Much of his seminal work was done in the 50s and 60s while at Columbia. "He was never a 'Get-one-more-paper-out' type of chemist," recalled a colleague. "He did the best of science, he was not driven by the ego that some of the other giants displayed." As to his personal character, Walling has been variously described as a deep thinker, a low-key, mischievous child, a founding father of free radical chemistry, a statesman in the world of science, a man with a quirky sense of humor, and the classic, gentleman-scholar. He once described himself as "an innocent bystander."

"I went into science because I wanted the freedom to do the work I wanted, to publish, to interact with other scientists. It is also a useful and respectable thing to do. My uncle was a Professor of Geology at Harvard and was looked up to by my family. I wasn't as much interested in contact with students as I was in understanding physical phenomena and explaining them to other people."

Walling's former students speak well of their days in his group and remain fond of this man. "He gave problems to his students, then we were off on our own," one recalled. "He was always available when you sought him out. But he did not lurk around seeking the latest results or looking over our shoulders as many of today's [academics] do. The projects were ours, not his. And he lit up like the sun when you got some results! It evoked great pleasure in him, and this excitement was transferred to us. He showed his emotions about science, not about the personal lives of the students."

Interestingly, several of his students independently described their professor as being "whimsical". Walling, according to Peter Wagner, one of his students, "was whimsical but a very serious scientist who was bemused by the intricacies of nature. When faced with an unusual result, he would say, 'What hath God wrought now?'" Walling liked being surprised by experiments not working as predicted. He had fun figuring out nature, and it was a contagious attitude that spread to his students.

His whimsy gave way to humor on occasion. Some of his students recalled, and could still describe, the outrageously garish ties he would wear to faculty and other professional meetings. One of his colleagues recalls seeing him " walking down the street in a raccoon-skin coat and a hat with an arrow pierced through, looking as if he were back in the days of the old West and had just survived an ambush where he had been shot through the head!" Another reminisced that "Cheves took it upon himself to mix the punch at the Department Christmas parties. It was a potent conglomeration of things. In an impish way, he tried to get everyone plastered!"

In turn, Walling reflects on how he saw himself as a teacher. "I was not a demanding research supervisor," Walling recently stated. "Students worked on their own. I was a better supervisor for people who themselves were quite good. Almost all of my students got their degrees."

Walling's passions are not worn on his shoulder, on display. "He did not jump up and down on the table to get his point across," remembers his colleague from Columbia, Gilbert Stork. "In that sense, he was really out of touch with the current mode of behavior! Cheves cared a lot about what he was interested in, from his science, to the Chemistry Department, to JACS (the Journal of the American Chemical Society). Walling provided the standards, the proper ethical standards where the discipline should be." Clearly, Walling was willing to take his time to better science.

In 1969, after 17 years at Columbia University, Cheves Walling, though no longer wearing the hat with the arrow through it, began to look seriously to the American West for his future. "My colleagues certainly thought I was crazy to move from Columbia to the University of Utah," he recalled. "But I was disillusioned with the student revolts [of the 1960s]. I'd been at Columbia a long time, knew how everything worked, wanted to try something new. No doubt I spent the early part of my career getting my reputation established. By 1969, I was 53 and could do what I wanted! And the west offered a very desirable geographic location (and lifestyle)." The years at Utah included further research in radical chemistry, brief forays into the nonclassical ion question and cold fusion, and Editorship of JACS.

In 1975 Cheves Walling took on the enormous responsibilities of Editor-in-Chief of the prestigious Journal of the American Chemical Society (JACS). Surely it was his innate curiosity and desire to share knowledge that made him an outstanding editor. For seven years Walling juggled the demands of publishing with his research at the University of Utah, sometimes sacrificing opportunities to conduct his own research in order to meet the demands of his publication. He also served for many years on the ACS's Committee on Professional Training, another voluntary disruption to his personal research.

Walling's statesman-like personality, his strength of character, his analytical, detached, and objective manner, and his mode of thinking made him a natural for the position at JACS. "The man lives by great principles to which he adheres without being bothered by current fashion or turns of fate. You agree with him or not, but you never have any sense of hidden agendas. You can see his honesty and strong beliefs. People think highly of him."

According to Peter Stang, a colleague at Utah, "Walling, together with Bob Parry (who had just moved from Michigan), and of course Henry Eyring, provided the foundation upon which the department's present prominence was built. Walling's high standard of excellence and gentlemanly demeanor served as the role model for young faculty as well as students." Besides his devotion to science, Walling loves the outdoors and in particular, the canyons and wilderness of Southern Utah. He was also an avid skier well into his 70s and would often go up for an afternoon of skiing after teaching his early morning class.

In 1986, the year of his 70th birthday, and just before he planned to retire, he offered to assist Dr. Stanley Pons, then Chairman of the Chemistry Department, and his collaborator, Dr. Martin Fleischmann, on the cold fusion experiments. "I was closing up shop and had spare time to get involved. I thought that they could use some support and help. I was the senior person in the Department." With those innocuous words Cheves Walling stepped into controversy.

Walling's involvement in the cold fusion controversy is discussed within the pages of this autobiography. Suffice it to say, ethical issues were certainly at the core of his excursion into the polemical field of cold fusion. "Walling felt that Pons and Fleischmann were not being treated with proper due respect, scientifically," explained a Utah colleague recently. "He felt quite strongly they should be considered innocent before being judged as guilty." But then the situation really heated up, with tremendous media coverage and international scientific scrutiny, criticism, secrecy regarding experimental data, threats of legal action, and involvement of tremendous power issues of prestige and funding. "At first, Pons and Fleischmann were pleased with my involvement," Walling recalls, "but as I became critical of things, Pons seemed to become somewhat paranoid. He could be a rather erratic character."

Now retired to his once summer retreat in New Hampshire, Walling reflected, "What matters the most to me in life? I am happy about my family. My wife has put up with me, basically giving up her own career. I found a career that I wanted to do and which would pay me to do it. [Much hearty laughter!!] And the best way to travel is as a distinguished visitor. [Even more laughter!!!]"

Cheves Walling is like a large river that seems calm and at rest when, in fact, it is moving quite rapidly beneath the surface, a strong current pulling it on ceaselessly. Walling has seen many milestones since the early days of his chemistry in the late 1930s, yet, like the river, he appears gentle, even serene, showing no signs of the currents that have affected his life.

JEFFREY I. SEEMAN
Philip Morris Research Center
Richmond, VA 23234

May 30, 1995

Fifty Years of Free Radicals

Cheves Walling

Early Life and Education

My Introduction to Chemistry

Late converts to chemistry are rare. I date my own interest to an encounter with a friend's chemistry set in the summer of 1925, just before I entered fifth grade. The charming color and odor of burning sulfur put a chemistry set on my Christmas wish list, and I was soon engaged in pyrotechnics and preparing hydrogen sulfide, sulfur dioxide, and chlorine for my friends to smell. Chemistry sets, such as still exist, have certainly become safer but less interesting. In the process we've removed much of their allure and, I'm afraid, lost a significant means of attracting young people into our profession.

My introduction to organic chemistry came a year or two later, when I read E. E. Slosson's *Creative Chemistry*,[1] with its account of Kekulé dreaming in front of his fire, Henry Perkin's discovery of mauve, and the subsequent development of synthetic dyes, plastics, and other marvels. I was particularly taken by structural formulas and still remember my discomfiture when, on trying to convert empirical formulas to structural formulas, I discovered isomers. I think I was in sixth grade at the time.

At this time I was growing up in Winnetka, Illinois, a suburb on Lake Michigan north of Chicago, where my family, consisting of my parents, two older brothers, and a sister, had moved a few years before I was born. To give some family background, which I can do because my paternal grandmother was an enthusiastic genealogist, and southerners are always full of family stories, all my forebears were in the country by 1776, having come, at various times, from England, Scotland, France, Germany, and Holland, and a number fought in the Revolutionary War. My father's family came west to Indiana and Kentucky from Maryland and Virginia around 1800 and included several physicians and one political figure, William Hayden English, a congressman from Indiana and unsuccessful candidate for vice president in 1880. My mother, Frederika C. Haskell, was from South Carolina, where her family had either settled originally or drifted by migration from New England. My rather unusual first name derives from a great-great-grandfather, Langdon Cheves, who had been speaker of the House of Representatives and president of the Bank of the United States in the early 1800s. Many of my mother's family fought (and some had died) for the Confederacy during the Civil War. Her father, Col. Alexander C. Haskell, had the final duty of surrendering Lee's cavalry to Grant at Appomattox.

Both my parents were relative liberals in a conservative community. My father, Willoughby G. Walling, and his older brother, William English Walling, had both attended the University of Chicago at a time when social and political reform was

The author at age five in an uncharacteristic pose.

very much in the air and had shaped their careers accordingly. As I was growing up my father was the president of the Morris Plan Bank of Chicago and a member of a number of charitable and public interest boards. The bank specialized in small low-cost loans to individuals and small businesses. By careful selection it kept defaults low and could charge low interest rates, in notable contrast to the present credit-card industry. My uncle English was more visionary—he was one of the founders of the National Association for the Advancement of Colored People (NAACP) and an active member of the Socialist Party before World War I. We lived in considerable comfort, even during the ensuing depression, but my father tried hard to impress on us children that success carried with it a real obligation to public service and to those who were less fortunate.

Why I was attracted to science is hard to explain because there have been no scientists in my direct family before or after me, and I know that, at first, my parents preferred me to choose medicine as my career. However, as early as I can remember, I was curious about and interested in the physical world around me. Immediately after World War I my father was vice president of the American Red Cross, and we lived in Washington, DC. I was so fascinated by the National Zoo that I constructed an imaginary duplicate in our yard that included elephants in the pergola and a Tasmanian devil under the woodshed. In those pretelevision days I was read to a good deal. As soon as I was able, I became an enthusiastic reader myself, with a taste for popular science, which I barely understood. The second film I ever saw was *The Lost World*, and I became intrigued by dinosaurs whose fossil bones I was then taken to see at the Field Museum in Chicago. Finally in the third grade, we studied the solar system complete with an imaginary space voyage to all of the planets. When I had learned to do compound multiplication, I remember trying to calculate how many miles were in a light year. All these things were interesting, but chemistry was the first branch of science I encountered in which I could do something rather than just read about it.

At the time of my introduction I was attending the North Shore Country Day School, a local private day school. It had an excellent college preparatory program but emphasized Latin, English, and history more than science, so I was largely self-taught in science. I soon discovered that the contents of my chemistry set could be amplified by interesting things like potassium chlorate and sulfuric acid from our local drug store, and I am glad to say that by the eighth grade I had scared myself out of making explosives after an incident that resulted in some burned fingers but no serious injury.

By high school my interests had broadened. Summers I attended a sailing camp on Cape Cod and became an enthusiastic small-boat sailor, a skill that has stood me in good stead at Gordon Research Conferences and elsewhere. There wasn't much chance for scientific experimentation in camp, but one year we wired a battery and Ford spark coil to the cabin doorknob, and I remember trying to stay awake until our counselor came back from his night off in order to close the switch.

In winters I spent much time building model airplanes, including some rocket-propelled models that enjoyed short but spectacular lives. I began reading Eddington and Jeans on cosmology, relativity, and quantum mechanics and have remained thoroughly confused about all three. I also became an avid reader of science fiction, although I've subsequently lost a good deal of my taste for it.

As a final item, in my senior year in high school, three classmates, Hobart Young, Jack Leslie, and Roderick Webster, and I discovered a 1923 Stanley steamer for sale for storage charges in a local garage. Fortunately it included a book of directions, so we pooled resources, bought it, and, after 2 days of hard work, were finally able to get up steam. A Stanley steamer had a fire-tube boiler that ran at 500 psi of steam pressure, a kerosene burner, a two-cylinder engine under the backseat, yards and yards of piping, dozens of valves, and many other fascinating oddities. During the spring it provided us with a thorough education in steam engineering and minor car repair. It also led me to make my first contribution to applied physics. We wanted to know how much power its engine developed, and I realized that if we knew the weight of the car, took it up to top speed, allowed it to coast to a stop, and plotted the rate at which it was slowing down against its speed, we could calculate the power delivered to the wheels for each speed along the way. The car would sustain about 35 mph with its burner going full blast, although it would go considerably faster for short

Spring 1933, our senior year in high school. Left to right: Hobart P. Young Jr., Jack Leslie, me, and Roderick Webster, who is behind the wheel. We are standing in front of our pride and joy, a 1923 Stanley Steamer. The steam-powered car was my introduction to applied thermodynamics.

spurts. After a few afternoons of getting up to speed and then coasting with a stopwatch and an eye on the speedometer, we were disappointed to conclude we could only develop about 25 horsepower. The car was still operating when we graduated, so that summer we set out in it for a tour of the West. This was my first trip across the Mississippi and my first introduction to the wonders of western scenery. I'm sorry to say that, after a number of implausible adventures, the glorious machine collapsed in Hayden, Colorado, and finished its days as a sheepherder's wagon. We still hold reunions over its career.

Despite these distractions, my interest in chemistry remained. We had moved my laboratory to Hobart Young's basement, where there was more room and gas for Bunsen burners. We also made the happy discovery that the Welch Scientific Company in Chicago not only sold chemicals, but practically gave away obsolete glassware. We soon had a splendid collection of flasks and condensers (unfortunately mostly soft glass) with which we studied fractional distillation by fermenting and distilling molasses. A more ambitious project to make ethanol by hydrating acetylene and then hydrogenating the acetaldehyde never got off the ground, as we barely got enough acetaldehyde to smell. A bit later we managed a rather appallingly successful experiment in smoke generation by injecting stannic chloride into the exhaust manifold of a Model-A Ford. It was so good we only did it once.

I think it was the applied and rather colorful nature of our experiments that led us both to decide to major in chemistry when we went off to room together as freshmen at Harvard University in 1933. This was despite the fact that neither of us had had a course in high school chemistry. For some arcane reason, at North Shore boys traditionally took physics and girls took chemistry because there wasn't enough room in the normal schedule for both subjects. Fortunately, I had learned enough on my own that I was able to get into the one-semester general chemistry course that presupposed an adequate high school course.

Undergraduate Study

At Harvard, my enthusiasm for chemistry was further strengthened by two excellent courses that I took as an undergraduate. The first, Chem 2, Introductory Organic, was taught by Louis Fieser, who had recently come to Harvard from Bryn Mawr College. Although the material was largely descriptive, Fieser loved the subject, and his enthusiasm made it exciting. Particularly memorable to me was his devotion to good laboratory technique and the glee with which he liked to demonstrate even such simple points as how to transfer a solution from one flask to the *inside* of another. Students in the course who had done well in the first semester were given a chance to work on individual projects in the second. In my case, this turned out to be an opportunity to work with C. H. (Hap) Fisher, the instructor in charge of the laboratory, and led to two short publications on some side-chain reactions of benzene derivatives,[2] my first appearance in the scientific literature. Hap himself left Harvard shortly thereafter for a successful career in the U.S. Department of Agriculture, and he and I have had pleasant encounters at meetings for many years.

The author at Harvard looking studious, about 1936. I'm afraid I had already started my pipe-smoking habit.

The second course, Chem 5, Advanced Organic, was taught by E. P. Kohler, then near the end of his career, and was more austere. There was more attention to how reactions occurred (I recall considerable discussion of 1,2 vs. 1,4 addition to carbonyl-conjugated olefins), and the assignments during reading period were papers on the new electronic theory developing in England and Carothers's recent work in polymer chemistry. Kohler was a polished lecturer but rather formal and reserved. Students stood rather in awe of him, and I only recall a few conversations with him.

I had much closer contact with Max Tishler, who was then the instructor in charge of the laboratory and due the next year to begin his brilliant career at Merck. I developed a lasting friendship with him. Max was a perfectionist who tried hard to convince me that care was as important as speed in laboratory work and that there was great merit in doing an experiment right the first time. Again I had a chance to do independent work, although this time, the project (an attempt to synthesize a highly substituted cyclobutadiene), rather to our disappointment, never led to anything conclusive.

Physical chemistry was taught by Henry Bent, whose son Henry Jr. was later to be a colleague of mine on the ACS Committee on Professional Training. Qualitative analysis, which I took as a freshman, was taught at a level of rigor that would not be accepted by students today. We were expected to commit the entire procedure to memory, and, at the end of the course, run through an entire analysis of a solution for the common cations and anions without the use of a text or notes. Although I've had little use for them and many are now obsolete, I still remember some of the procedures. The real value of the course was that I was forced, for the first time, to rely entirely on my own observations. I was introduced to the difference between neat statements in the text and equivocal observations at the bench-top. Is that little bit of stuff really a precipitate or not?

Quantitative analysis was taught by G. P. Baxter, one of the last workers to still determine atomic weights by classical analytical methods. One of his graduate students at the time, as I recall, was redetermining the atomic weights of krypton and xenon by laboriously separating a mixture by repeated partial evaporation from one glass bulb to another and periodically determining the density of the gas. This sort of thing was not my dish, and I regarded getting through the course chiefly as a test of my commitment to my chosen field. Other courses I took I considered rather a mixed bag, but I remember with pleasure a rigorous and demanding course in European history and an introduction to geology which has enhanced my interest in scenery and led me to a lifelong hobby of mineral collecting.

As for my fellow chemistry majors, then as now most were heading for medical careers, a handful went on for a Ph.D., but Saul G. Cohen, later professor and dean at Brandeis University, where he played an important part in setting up their graduate program in science, is the only classmate I know that ended up in chemical academia.

Harvard, as I recall it in those days, was very much of a class-conscious society. There was an elite, mostly from eastern

This was taken at French Lick, Indiana, where I had gone on a trip with my father during spring vacation, 1937. It shows I had other interests besides chemistry.

preparatory schools, largely in college for the social life and to cement connections that would be valuable to them later. At the other end, there were students, largely commuters, who were concerned primarily with what Harvard had to offer in terms of education and training for a profession. For this group, a degree offered a chance to move into a wider world. Finally, there was a large middle class, with rather mixed aspirations, to which I and most of my friends belonged. I enjoyed my 4 years there, which were certainly not devoted exclusively to chemistry. I learned to ski and, since Harvard was notably non-coeducational, spent considerable time driving to women's colleges. I have remained a loyal alumnus, although I will admit that I only occasionally see former classmates. My connections have been chiefly with the chemistry department, where I served a term on their visiting committee.

I might add that my old friend and roommate, Hobart Young, also went on into chemistry, working for Armour and Company, where he, for a time, ran the plant making chemicals from fatty acids. He finally retired as a vice-president of the Signode Corporation, where he developed the manufacture of plastic packaging strap, steel strapping having been Signode's major product. The Wallings and the Youngs have remained in touch and have traveled together considerably.

Preparation for Chemistry Career: Free Radicals Appear

By the spring of my senior year, I was firmly committed to a career in chemistry, but my goals were certainly fuzzy. I was still captivated by the qualitative pleasures of synthesis and the preparation of possibly new compounds and less taken by the sort of quantitative measurements done in physical chemistry. The question of just how I was going to practice my chosen profession was best deferred by going to graduate school, which, in any case, would be a requirement if I planned to teach.

My father was an enthusiastic alumnus of the University of Chicago and a friend and admirer of Robert M. Hutchins, then the university president. An inquiry at Harvard indicated that Chicago had a competent chemistry department and that (according to Max Tishler) the most interesting organic chemistry was being done by Morris Kharasch, who was then in the midst of his studies of the peroxide effect in the addition of hydrogen bromide to olefins. Accordingly, during my Easter vacation, my father arranged an appointment for me with Henry Gale, a friend of his, who was then, I believe, the dean of science. I visited the campus, and the dean sent me to the chemistry department, where I met several staff members including Kharasch, whom I found drinking coffee in the faculty lounge. Although I had been offered a teaching assistantship at Harvard by Louis Fieser, both my family and I could see advantages in continuing my education nearer home, and so I decided that Chicago was where I would go.

In the summer of 1937, after my graduation from Harvard, I spent 2 months touring much of Western Europe with three of my classmates. It was a happy-go-lucky trip only slightly marred by our impression of Nazi Germany. I had been there 2 years before, and the military activity and ominous atmosphere that had built up in the interim was frightening. As tourists, however, we were welcomed and well treated even when we blundered through the gates of an S.S. camp one evening in the dark. As a more gruesome touch, we accidentally spent one night in the now infamous village of Dachau outside of Munich. When we mentioned this to Germans later it turned out to be a complete conversation stopper. It wasn't until much later that we fully realized why.

When it came to enrolling at Chicago, procedures were much less elaborate in those days, and I simply turned up in September with my Harvard diploma and a check for tuition and asked Kharasch if I might join his group.

For the field I was inadvertently entering, 1937 was a landmark year. Free radicals of course first entered organic chemistry in 1900 with Gomberg's preparation and identification of triphenylmethyl,[3] but the chemistry and properties of such "stable" or "persistent" species had remained largely a chemical curiosity. The idea that free radicals could be transient intermediates in ordinary chemical reactions had come from quite a different source: the study by physical chemists of the high-temperature and photochemical reactions of small molecules in the gas phase. By the mid-1930s, transient radicals (commonly detected by their ability to remove thin metal mirrors) were well recognized as intermediates in a variety of such reactions, and what was known was well summarized by Rice and Rice in their book, *The Aliphatic Free Radicals.*[4]

Extension of these ideas to liquid-phase reactions near room temperature had come more slowly, perhaps because physical and organic chemists paid little attention to each other. Mechanistic ideas were dominated at the time by the developing electronic theory, with its emphasis on electron pairs and the dichotomy of electrophilic and nucleophilic reagents. At first there were really only tantalizing hints: Staudinger's 1920 suggestions that the ends of long polymer chains might be "free valences,"[5] Michaelis's studies of semiquinones,[6] Bäckström's chain mechanism for benzaldehyde oxidation,[7] and Haber and Willstätter's suggestion of radical mechanisms for several reactions[8] are examples.

The year 1937 though was marked by three very significant publications. First was a review by Hey and Waters,[9] rationalizing in terms of free radical intermediates the puzzling formation (and isomer distribution) of biaryls formed when benzoyl peroxide or aromatic diazo derivatives were decomposed in aromatic solvents and proposing radical intermediates in several other reactions, including the suggestion that Kharasch's abnormal addition of hydrogen bromide to olefins might involve bromine atom intermediates. The second was Kharasch's own formulation of a bromine atom chain mechanism for the abnormal addition in essentially its present form.[10] The third was a prescient paper by P. J. Flory[11] analyzing vinyl polymerization as a radical chain reaction and laying down much of the framework into which much subsequent work has been fitted.

Graduate Work at Chicago, 1937–1939

Most of this background was, of course, unknown to me in the fall of 1937. In discussing possible research problems with Kharasch, I remember him saying that anyone could synthesize new compounds but the interesting question was how reactions worked. After discussing several possibilities, we agreed that the problem on which I was to work was to determine the direction of addition of hydrogen bromide to *cis*- and *trans*-2-pentene and how this addition might vary with reaction conditions.

This problem had a tangled history. Kharasch had been an early and rather unorthodox exponent of the electronic theory. He had been interpreting the course of chemical reactions in terms of the relative electronegativities of substituent groups and had endeavored to determine these electronegativities by studying the direction of acid-catalyzed cleavage of dialkylmercury compounds, R_1HgR_2.[12] In these terms, the direction of addition to 2-pentene depended on the relative electronegativities of methyl and ethyl groups. The trouble was that different laboratories reported contradictory results.

This confusing situation had led Kharasch, in a 1928 review,[13] to propose that 2-pentene could exist as two isomers with different electronic structures. In one, an electron pair was localized at one end of the double bond, $C_2H_5\ddot{C}H{-}CHCH_3$, and at the other end in the other isomer, $C_2H_5CH{-}\ddot{C}HCH_3$. The then current electronic theory, in turn, predicted that the first isomer would add HBr (polarized as H^+Br^-) to give 2-bromopentane, and the second, 3-bromopentane. The possibility of such isomers, known as "electromers" had some currency at the time. With the development of quantum theory, however, it became evident that such structures would either represent resonance forms of the same molecule or a ground state and an excited state. In neither case would they be separable products. Similar contradictions had been observed in the addition of HBr to allyl bromide, and it was to test this hypothesis of electromers that Frank Mayo undertook the study that led eventually to the discovery of the peroxide effect.[14,15]

My goal was thus to reexamine the original system in terms of what was now known about the peroxide effect, newly interpreted as a bromine atom chain, and its competition with the usual ionic or polar reaction. The results, although they laid an old ghost to rest, were somewhat anticlimactic. Under all conditions, both *cis*- and *trans*-2-pentene gave, within experimental error, an equimolecular mixture of 2- and 3-bromopentane (equation 1). Similar results were obtained with HCl.[16]

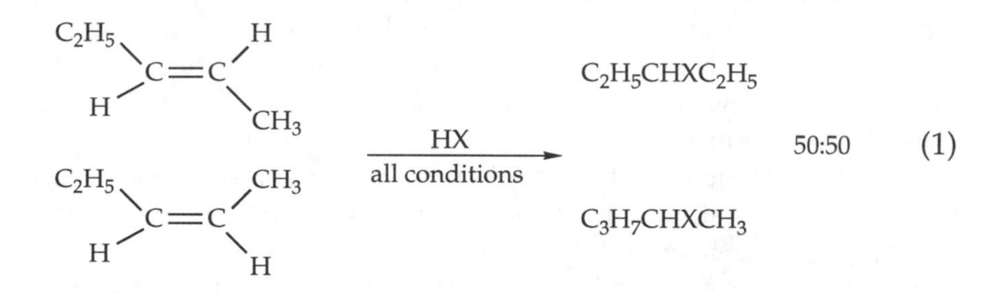

The work is most interesting now as a reflection of the state of chemical techniques at the time. Besides vacuum lines for carrying out reactions in the absence of air, Kharasch's research group had two "sophisticated" chemical instruments: a Podbelniak column (a fractional distillation setup with a 1.5-m-long, vacuum-jacketed column containing a wire spiral) through which starting materials and products were laboriously fractionated and an Abbé refractometer. Although 2- and 3-bromopentane were inseparable by distillation, their indices of refraction differed by 30 units in the fourth decimal place, and it was on this basis (plus occasional conversion to solid derivatives and matching melting points against melting-point phase diagrams for known mixtures) that product compositions were deduced. Because the pentenes themselves had to be synthesized by several steps, what would now be a trivial problem took me over a year to complete.

This work, together with two shorter projects, made up my thesis. One of these smaller projects, the addition of hydrogen bromide to 2-butyne,[17] I had devised as a quick way of demonstrating a peroxide effect in an addition to a nonterminal double bond, because what was necessarily the first product, 2-

bromo-2-butene, had a double bond that was both quite unsymmetric and somewhat deactivated toward normal addition. The experiments were successful in that, in the presence of peroxides and light, the normal product, 2,2-dibromobutane, was replaced by the abnormal product, 2,3-dibromobutane. The system provided a bonus in that the 2,3 dibromobutane was obtained solely as the racemic isomer. Although we reported this fact without comment, Goering, 20 years later, showed that this was the consequence of two successive *trans* additions,[18] one of the first examples of "bromine bridging" (equation 2).

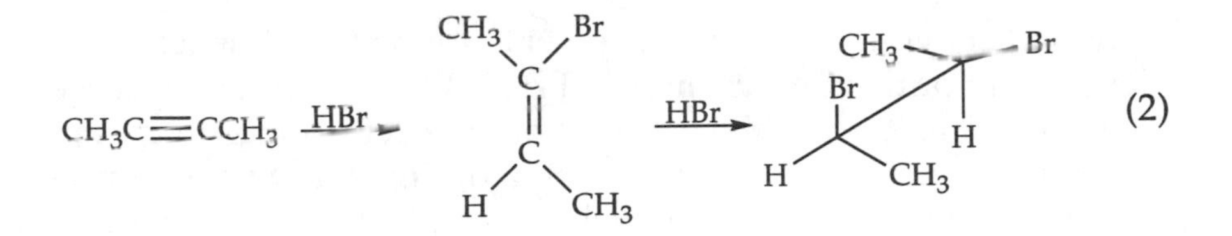

Kharasch's Research Group

At the time I was at Chicago, 1937–1939, Kharasch had quite a large research group. Most of his collaborators were working on various aspects of radical chain additions and substitutions, but Kharasch was also beginning his investigations of peroxide chemistry and the effect of trace metals on Grignard reactions. Toward the end of my time there, I did some experiments with benzoyl peroxide, and I recall declining to consider studying acetyl peroxide after learning about its explosive properties.

Kharasch was a busy man with extensive consulting activities, and although he treated us all to an annual Chinese dinner, we students saw him chiefly at occasional group seminars at which students reviewed areas of interest to the group's research. Later I came to know him better and found him very cordial whenever I visited Chicago. Most of my impressions come from this later period, but he was clearly one of the most original and independent organic chemists of his time. His independence led him into considerable controversy with the more synthetically oriented "establishment," but his papers, published between 1937 and his death in 1957, still contain nuggets of data and ideas waiting to be exploited.

The day-to-day direction of Kharasch's group was in the hands of Frank R. Mayo, who had received his Ph.D. with Kharasch in 1931 for work leading to the discovery of the peroxide effect. After a stint with DuPont, Mayo returned to Chicago as an instructor. Our association at that time developed into a lifelong friendship and, a bit later, a highly productive collaboration.

Important Influences

Two other young faculty members strongly influenced my developing outlook on chemistry: Frank Westheimer and George Wheland. Frank Westheimer taught an excellent course in chemical kinetics and was, I found, a penetrating questioner during oral examinations. Frank had held a postdoctoral fellowship at Columbia University with Louis Hammett, whom he greatly admired. It was from Frank that I first heard of acidity functions. His research included a collaboration with J. G. Kirkwood on the quantitative calculation of electrostatic and steric effects. The latter project, involving the calculation of the barrier to racemization in a hindered biphenyl, was the first successful application of what is now known as "molecular mechanics."

George Wheland came from Linus Pauling's laboratory. In his courses, he outlined the newly developing ideas of the effects of electron distribution and resonance on organic structures and reactions, ideas he was to elaborate in his very influential books, *The Theory of Resonance* and *Resonance in Organic Chemistry*. I recall a good deal of initial skepticism about these ideas on the part of students, but he, Frank, and I used to eat lunch together, and I came to appreciate his scholarship.

Finally, it was at Chicago that I first met H. C. Brown. His Ph.D. research was in inorganic chemistry, but he joined Kharasch's group as a postdoctoral fellow in 1939 and did important work on radical chain halogenations.[19] I recall him as a productive and ingenious experimentalist, already sharpening his skills as a vigorous advocate of his own ideas.

In the spring of 1939, Kharasch invited Frank Mayo and me to collaborate on a *Chemical Reviews* article on the peroxide

At the 2nd Reaction Mechanisms Conference at Colby Junior College, New London, New Hampshire, in 1948. Left to right: Frank Westheimer, Morris Kharasch, and Frank Mayo. At this time Westheimer was still at Chicago and had not yet moved to Harvard

effect. I volunteered to round up the pertinent literature and to produce a preliminary draft. This was a very valuable and productive experience for me. It familiarized me with the literature, and because Frank was a meticulous writer and critic, I learned much about clarity, accuracy, and succinctness in writing. I had also been reading Rice and Rice's book,[4] which pointed out the importance of bond strengths and energetics in the elementary steps in radical reactions, and I was able to suggest the reason why hydrogen bromide, of all the halogen acids, was uniquely able to take part in a radical chain. With HBr alone were both chain propagation steps strongly exothermic processes (equations 3 and 4).

$$Br\cdot + CH_2{=}CHR \longrightarrow BrCH_2\dot{C}HR \quad \Delta H \cong -5 \text{ kcal} \tag{3}$$

$$BrCH_2\dot{C}HR + HBr \longrightarrow BrCH_2CH_2R + Br\cdot \quad \Delta H \cong -11 \text{ kcal} \tag{4}$$

A similar argument based on energetics also accounted for the direction of addition. Both of these ideas were incorporated in the review. When I left Chicago, I left the draft with Frank for final revision and polishing. The review appeared in 1940[20] without Kharasch's name, because he agreed with Frank that we had done the actual writing.

Incidentally, although the values of ΔH in equations 3 and 4 were only known roughly in 1939, the fact that ΔHs for the individual steps in radical reactions can be reliably calculated if relevant bond dissociation energies are known has proved to be a powerful guiding principle in the development of free radical chemistry and is now discussed even in elementary texts.

Consequently, much effort has gone into the measurement of bond dissociation energies. Reliable values for simple molecules are now generally available, but measurements on larger species with multiple substituents still present a formidable problem.

With bond dissociation energies known, minimum values for activation energies can be estimated, and postulated processes that would be too slow can be ruled out. Further, by using semiempirical values for entropies of activation and excess energies required to surmount activation barriers, rate constants can sometimes be estimated as well.

Starting Out in Industry

E. I. du Pont de Nemours and Company, 1939–1943

My other problem, as the completion of my Ph.D. approached, was finding a job. My preference would have been for an academic position, and I managed to be the runner-up for a position at the University of Rochester. Fortunately, I was also interviewed by DuPont, and they offered me a position at their Jackson Laboratories, which I accepted. I find it amusing that the successful applicant for the Rochester position was T. L. Cairns. We subsequently switched careers, because he went on to DuPont to become a very well known industrial chemist and finally their director of research.

In September 1939, I left for the East, stopping on the way to attend the American Chemical Society meeting in Boston, where I gave a paper on some of my work, including the argument for the uniquely favorable energetics of the HBr reaction. As I recall, the audience was even smaller than it might have been; many of the attendees had been stricken by food poisoning at a now-defunct hotel in Swampscott the night before.

DuPont was an abrupt change in many ways. The Jackson Laboratories were in Deepwater, New Jersey, across the Delaware River from Wilmington. They were a part of an enormous chemical plant, "The Dye Works", which produced dyes and a variety of organic products from neoprene to Freon, emitting its effluvia into the air and, in brightly colored streams, into the river. There was no bridge across the river in those days, and those of us who lived in Wilmington got to work via a company ferry known affectionately as the "cattle boat", which left Wilmington promptly every morning a little after 7:00 A.M. As the war intensified there was increasing naval traffic in the river.

One morning a badly damaged British battleship emerged from the fog, down at the bow, but with an enormous ensign flying defiantly from the stern, on its way to the Philadelphia navy yard for repairs.

Initially I worked on azo dyes. Every fiber has its own peculiar dying properties, and I worked chiefly on new metal-chelate dyes for nylon, which had just come on the market and was proving difficult to dye with standard dyes. Azo dye synthesis is all aqueous chemistry and was carried out in rows of stirred beakers. Reactions were followed and controlled by a crude form of paper chromatography, a technique of which I had never heard although it had been known for years among dye chemists.

The laboratory was full of bottles of chemicals like benzidine, which are now considered acutely carcinogenic. Actually, by standards of the time DuPont was very safety-conscious. In the plant where exposure was greater, considerable precautions were taken. Workers wore protective clothing and were given periodic physical examinations. If one developed symptoms of some adverse reaction the policy was to let you off work until you recovered and then assign you to a different job.

In my case I came down with acute chemical dermatitis after a year (recalling what I had been working with, I think I was an early discoverer of the physiological effects of dioxins) and was transferred to research on fuel- and lubricating-oil additives. Here I had no more trouble, although I had two fires with organophosphorus compounds, which gave me a healthy respect for them. For a time I shared a laboratory with C. J. Pedersen, who subsequently became famous as the discoverer of crown ethers, for which he shared a Nobel prize. I recall him as a quiet, rather oriental-looking, bachelor (his father was Norwegian and his mother was Korean); an agreeable colleague; and a skillful, careful research worker, who had developed some gasoline additives of considerable commercial value to the company. I'm delighted that his skill and persistence have been so well recognized.

During the 3 years I spent at DuPont, my efforts led to a few patents but nothing that ever reached the marketplace. I learned a good deal about industrial processes, patent law, and the many hurdles a successful product must surmount. The work

Our growing family in Montclair, New Jersey, while I was at U.S. Rubber in 1946. Left to right: Jane, Rosalind (age 3), Cheves (quite new), and Hazel (age 5). I think the picture was taken for the company magazine to show the happy home life of their employees. (Photo taken by Samuel W. Vandivert.)

was largely empirical ("Edisonian research" was a common phrase), however, and I found it increasingly frustrating.

On the other hand, my nonchemical life prospered. In 1940, I married my first (and only) wife, Jane Wilson, a Vassar graduate whom I had met in Winnetka while I was in graduate school. In 1941, our first daughter was born to be followed, at various intervals, by three more daughters and a son. All have married at least once, and the family now includes ten grand- and two great-grandchildren. Jane has stuck by me through several moves, somewhat compensated by the chance to travel extensively. Besides raising the family and acting as my social director, she has pursued an active career of her own, partly in voluntary activities and partly as a professional social worker, having taken an M.S.W. degree at the University of Utah in 1972. We celebrated our 50th wedding anniversary in 1990 and now take care of each other in retirement.

Polymer Chemistry: The U.S. Rubber Company, 1943–1949

Meanwhile, world events had overtaken the leisurely pace of academic science and ordinary industrial research. World War II began in 1939. Japan attacked Pearl Harbor in 1941, and scientists were largely drawn into war-related work. The Japanese occupation of Southeast Asia and the East Indies cut off the Allies' supply of natural rubber, and a large synthetic rubber industry had to be created largely from scratch. Polymer chemistry suddenly became vitally important, and because much of it was free radical chemistry, radical chemistry became important too.

Among the organizations caught up in this whirl was the U.S. Rubber Company (subsequently Uniroyal and now essentially vanished through a series of reorganizations, sales, and mergers). In addition to operating one of the major synthetic rubber plants and carrying on the day-to-day production of war materials, their management decided to set up a long-range research program in fundamental polymer chemistry. The chief proponents of this plan were W. A. Gibbons, the director of research at their general laboratories, and R. T. Anderson, an associate director, who had come from Massachusetts Institute of Technology (MIT) and subsequently went on to Celanese.

Several groups were set up, in particular, one in physical chemistry, headed by R. W. Ewart (of whom more later), and one in organic chemistry, for which they recruited my old friend Frank Mayo. The general laboratories were in the upper floors of an old silk mill in Passaic, New Jersey, across the street from a U.S. Rubber plant that made everything from boots to garden hoses. Frank set up shop there in the fall of 1942 and, shortly thereafter, asked me to become a member of his group. The idea was attractive because I liked and admired Frank; the research sounded new, interesting, and more closely related to the war effort; and I felt the need for a change. Negotiations proved to be a bit delicate, because mutual nonhiring agreements between companies were still legal and popular. I duly gave DuPont notice and, after a week's skiing vacation with my wife, joined U.S. Rubber in February 1943. Passaic was hardly a garden spot, so we rented and later bought a house in Upper Montclair about 6 miles away. Connecting public transportation was poor and gas-rationing was in force, so I took up riding a bicycle—a pleasant experience in good weather and an excellent recreation.

I found myself in a new and very exciting world. Frank had already analyzed two important problems in vinyl polymerization: (1) the effect of chain transfer on polymer molecular weight and (2) the problem of predicting copolymer compositions in terms of simple, easily evaluated ratios of rate constants. He was gathering data to verify his analyses. The concepts were quite simple and followed what were, even then, recognized as the basic tenets of vinyl polymerization.

Polymerization starts by addition of some sort of initiator fragment to a monomer molecule to yield an active center. The polymer chain is then propagated by successive additions of monomer to the active center (in radical polymerization, a free radical) until the center is destroyed by some termination process. Essentially, the polymer arises from the propagation step (equation 5):

$$\sim\!M\cdot + M \xrightarrow{k_p} \sim\!M\cdot \qquad (5)$$

with the polymerization rate constant, k_p, assumed to be independent of chain length. Chain transfer, which had been considered as a possibility by Flory,[11] involves the reaction of ~M·, the growing chain, with some solvent or other molecule X–Y (equation 6; k_{tr} is the rate constant for the transfer reaction)

$$\sim\!M\cdot + X\text{–}Y \xrightarrow{k_{tr}} \sim\!MX + Y\cdot \qquad (6)$$

to yield a fragment which then restarts the chain (equation 7)

$$Y\cdot + M \longrightarrow Y\text{–}M\cdot \longrightarrow \sim\!M\cdot \qquad (7)$$

Each act of transfer thus cuts an otherwise long physical chain in two without significantly affecting polymerization rate, and polymer molecular weight can be related to the transfer constant, $C_{tr} = k_{tr}/k_p$.[21]

For copolymerization involving two monomers, M_1 and M_2, Mayo introduced the assumption that the reactive properties of the growing chain depended solely on the monomer unit carrying the active center so that chain propagation should involve four rate constants (Scheme 1). Scheme 1 leads to the now-familiar copolymerization equation in which compositions of polymer and feed are related by two monomer reactivity ratios, $r_1 = k_{11}/k_{12}$ and $r_2 = k_{22}/k_{21}$, characterizing the selective properties of $\sim M_1\cdot$ and $\sim M_2\cdot$.[22]

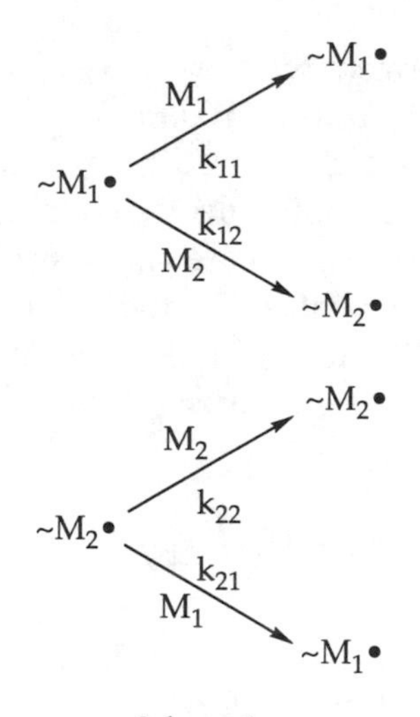

Scheme 1.

Both analyses were straightforward and were, in fact, carried out independently by other investigators.[23] The great advantage at the U.S. Rubber Company was that we had an adequate and enthusiastic group able to follow up the theory by experiment. Because we were backed by a strong analytical laboratory, we were able to examine a large number of systems and to produce quantities of valuable data from which important generalizations could be deduced.

Taken at a farewell luncheon before I left U.S. Rubber in 1949. I can't identify everyone, but I am the fourth from the left followed by Max Matheson, Ken Doak, Fred Lewis, and Bob Gregg. The last three were all members of Frank Mayo's research group.

While I was there, the Mayo group fluctuated somewhat in size and composition, but the three stalwarts were F. M. Lewis, R. G. Gregg, and K. W. Doak. Fred Lewis's contributions, I have always felt, were particularly noteworthy. He was an ingenious experimentalist and a fertile source of ideas. He did most of the early work on copolymerization, much of it while working evenings and weekends toward his Ph.D. at New York University.

I also contributed to the fluctuation in that I was absent for 6 months myself in 1945, on loan to the Office of Scientific Research and Development (OSRD) in Washington, DC, the result of a chance encounter between my wife and an old friend whose husband was the chief legal aide to OSRD. My position, technical aide in the antimalarial drug program, had nothing to do with polymers or radicals, but it gave me an interesting view of wartime Washington. More important, because most of the synthesis of potential new drugs was being carried out in university laboratories that I was expected to visit, it led me to meeting a large number of organic chemists, spread all over the country.

My last extensive trip was in August 1945. The night before I had borrowed a prerelease copy of the Smyth Report, the now largely forgotten document that described the whole Manhattan Project (which had produced the atomic bombs) in remarkable detail. I sat up most the night reading it. It provided quite a topic of conversation during my trip. Word that the war had ended reached the train as it went through Hagerstown, Maryland. The conductor promptly closed the bar, so our celebration was more restrained than it might have been.

To look back at a broader picture, the explosive growth in the understanding of polymer and radical chemistry that was taking place in this period really came from three sources: the unexploited background of small-molecule chemistry that was available, the number of talented scientists who were drawn into the field, and the relatively free exchange of information that was going on between different research groups.

The nerve center for this exchange in the Northeast was certainly "Brooklyn Poly" (now the Polytechnic Institute of New York). Herman Mark had started a polymer institute there and was busy educating American chemists on European polymer chemistry. Mark's contributions to American polymer chemistry can hardly be overstated, but most notable to us at the time was a regular Saturday morning seminar (often lasting all day), which we frequented and which chemists from all over the East Coast attended to exchange ideas. One who came quite frequently and whose interests clearly paralleled my own was Paul D. Bartlett from Harvard. Our long association began during this period.

In addition, U.S. Rubber made extensive use of academic consultants. At one time or another, W. D. Harkins, P. P. Debye, W. G. Young, Saul Winstein, Byron Riegel, and

At a Gordon Research Conference in New Hampton, New Hampshire, 1971. Saul Winstein had started a custom of wearing loud Hawaiian shirts to conferences, and this was my ultimate response, a souvenir from Antigua. Left to right: Rolf Huisgen, Jack Roberts, Edith Roberts, me, and Stan Cristol.

Morris Kharasch were all visitors to the laboratories, and we all profited from our discussions with them. I saw Young and Winstein frequently in later years, and Debye was particularly impressive. From time to time, he gave us elegant little lectures on such topics as light scattering by polymer solutions and the concentration dependence of viscosity, which were so crystal clear that, for a time, I would think I understood what he was talking about.

Structure and Reactivity in Radical Chemistry

My own research involved several aspects of polymerization kinetics, including the application of Flory's statistical approach to gel formation[24] to polymerizations involving bifunctional monomers for example, methyl methacrylate containing ethylene dimethacrylate. According to the theory, solutions gel when infinitely large networks of cross-linked polymers form. The point at which gel formation should take place is predicted by a simple theory. In practice, dilute systems fail to gel as early as predicted by the theory. This was my introduction to the importance of intramolecular reactions in radical systems. Under dilute conditions, reactions that should have led to cross-linked networks were being wasted by the reaction of molecules with themselves.[25] Increasingly, though, my attention was drawn to copolymerization and the rich source of data it provided on the relation between structure and reactivity. Most of my work in 1943–1949, the years I spent at U.S. Rubber, was in this area.

As the data on new monomer pairs accumulated, a pattern emerged, showing that three factors were important. The first factor was simply the familiar parallel between energetics and rate. Olefins that gave rise to resonance-stabilized radicals, for example, styrene or acrylonitrile, were more reactive than olefins giving localized radicals, such as vinyl acetate, and conversely, the resulting stabilized radicals were less reactive.

The second factor arose from the observation that 1,2-disubstituted olefins were, in general, less reactive than those with substitution at only one end of the double bond. This fact, together with the generally accepted view that vinyl polymers had head-to-tail structures and the known direction of addition of small radicals (e.g. Br•) to olefins, implied that steric hindrance was important in radical additions.

The third finding was entirely unexpected and was probably the most important discovery that came from the copolymerization work. In the very first paper by Mayo and Lewis,[22] in which monomer reactivity ratios for the styrene–methyl methacrylate system were reported, it was found that each radical preferred to react with the *other* monomer by a factor of 2:1. In short, the two monomer units tended to *alternate* along the polymer chain. More data showed that this alternating effect was quite general and paralleled *differences* in monomer polarity, that is, the effect was greatest when one monomer carried electron-accepting groups and the other carried electron-donating groups. Thus, alternation was the result of a polar effect. The practical consequences of this finding were striking.

Because many monomer pairs tended to alternate, it was relatively easy to prepare a wide variety of copolymers containing significant amounts of both monomers, and so, the variety of products available from radical polymerization were greatly increased. Many important commercial plastics are such copolymers, for example, GR-N or nitrile rubber, an oil-resistant butadiene–acrylonitrile copolymer, and high-impact polystyrene, an acrylonitrile–styrene copolymer.

No similar phenomenon was found in polymerizations involving carbanion or carbocation active centers. With such active centers, copolymers can be produced efficiently only from monomers of very similar reactivity, and so a smaller range of products are possible. Copolymer compositions, it turns out, provide a useful tool for distinguishing radical and nonradical active centers in polymerization reactions. As an example, mixtures of styrene and methyl methacrylate, in the presence of radical sources, give an initial polymer of essentially the same composition as the feed. Cationic polymerization, for example, with $SnCl_4$, gives almost pure polystyrene, whereas carbanionic polymerization, such as that induced by sodium triphenylmethide, gives initially poly(methyl methacrylate).[26] This finding, incidentally, was one of the first investigations I made at U.S. Rubber, although publication was delayed for commercial reasons.

The explanation of this polar effect was more challenging. One factor could well be simple electrostatic interaction between permanent and induced dipoles in the radical and the monomer. Such a model provided the basis for the "Q and e" scheme of Price and Alfrey[27] for predicting monomer reactivity ratios. However, in strongly alternating systems such as styrene–maleic anhydride, which gives an almost perfectly alternating copolymer, the effect was so large that some further explanation seemed in order.

An answer that emerged was that, in such systems, the barriers to reaction might be further lowered by actual electron transfer in the transition states, which would accordingly become resonance hybrids involving a number of basic structures. For styrene–maleic anhydride, for example, the basic structures are represented by **1** and **2**.

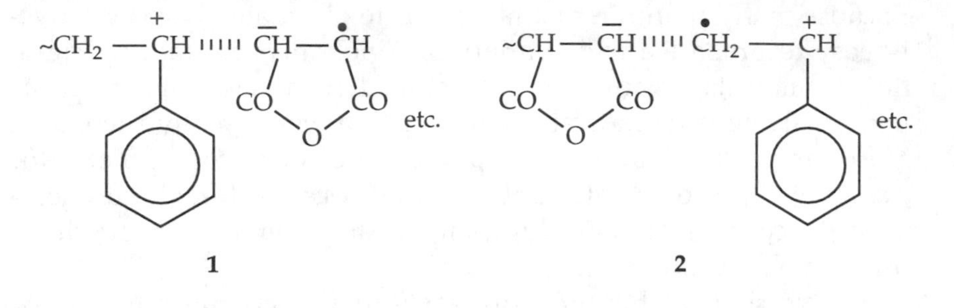

One line of argument was based on Hammett $\sigma-\rho$ correlations, which we observed with substituted styrenes. Hammett's ground-breaking book, *Physical Organic Chemistry,*[28] was just beginning to have its profound influence on chemical thinking in the United States, and the use of linear free energy relations was just beginning to be appreciated. Actually, it so happened that I had been intrigued by them for some time. My senior year at Harvard I had subscribed to the *Journal of the American Chemical Society.* The first paper I read in the first issue I received was Hammett's exposition of his famous equation[29], and I never forgot it.

Application of the Hammett equation to radical reactions as a means of separating electronic effects from other effects had not previously been tried, and our results were mixed. Reactivities of substituted styrenes with the styryl radical gave a good Hammett plot with a small positive slope.[30] On the other hand, α-methylstyrenes with electron-supplying groups (–R, –OR, etc.) showed unexpectedly high reactivities with electron-poor radicals, such as that from maleic anhydride[31] and the thiyl radical[32] from thioglycollic acid. Brown's σ^+ constants[33] had not yet been invented, but we noted that there were similar failures of Hammett plots in solvolyses leading to benzylic cations, supporting our idea of actual positive charges in our transition states as well. Furthermore, charge-transfer models for molecular complexes had become commonly accepted, and we noted a strong parallel between our results and the colors and stabilities of complexes formed between maleic anhydride monomer and our substituted styrenes.

The chronology of these ideas is (and was) rather obscure. Our full exposition was published in 1948,[30–32] but the charge-transfer concept had been noted earlier by Bartlett and No-

Frank Mayo (standing) and Paul Bartlett, taken at a meeting at the Stanford Research Institute in 1968. Both were among my closest friends and had a great influence on my career.

zaki.[34] The whole matter had been discussed a number of times at meetings and seminars.[35] When Mayo and I published a review of copolymerization in 1950,[36] we proposed that polar effects and electron transfer in transition states should be general phenomena in determining reactivities in radical reactions, that is, that reactions should occur with particular ease between electron-rich radicals and electron-poor substrates and vice versa, and we noted a few apparent examples.

Although a quantitative general theory of the polar effect is still lacking, our prediction was certainly qualitatively correct. Enhanced rates of radical reactions between radicals and substrates of different polarity are a common phenomena, and polar effects have been observed (or rediscovered) in many laboratories. In short, the idea that radicals may have electrophilic or nucleophilic properties (behaving as electron acceptors or electron donors, respectively) is now generally accepted and has proved very useful in devising highly reactive systems for use in synthesis. It is interesting that the converse of this polar effect in radical reactions, involving the postulation of radical structures for intermediates or transition states in what had previously been considered simple polar additions or displacements, is currently attracting attention under the name of single-electron-transfer (SET) reactions.

Other Developments in Free Radical Chemistry

These structure–reactivity relationships were only a small part of the new knowledge of radical chemistry that came out of polymer research during and immediately after the war. As part of the interest in initiator systems, the chemistry of organic peroxides began to unfold, including induced decompositions and competitions between polar and radical decomposition paths. The mechanisms of inhibitor reactions were unraveled, a particularly interesting example being the process Bartlett and Altschul[37] termed "degradative chain transfer," which explained why ethyl acrylate polymerizes readily to a high-molecular-weight product, whereas its isomer, allyl acetate, reacts only slowly to give low-molecular-weight products. With allyl acetate, a minor side reaction of chain transfer with allylic hydrogens gives an allylic radical too unreactive to continue the chain. Similar side reactions are now recognized as accounting for many failures of otherwise promising radical chains.

The first absolute rate constants for chain propagation and termination in polymerizations became available, measured most successfully by the rotating-sector technique.[38] These data not only established the time scale for the kinetic chains involved but, combined with copolymerization relative reactivities, also gave the first information on the relative rates of reaction of different radicals with the same substrate and showed that the relative rates were governed by much the same factors as had been deduced from copolymerization data.

Finally, a plausible explanation of the serendipitous nature of emulsion polymerization emerged. Emulsion polymerizations employ an aqueous emulsion of monomer in water plus soap (or other emulsifying agent), a radical initiator, and perhaps other additives. The bulk of the synthetic rubber (a styrene–butadiene copolymer) manufactured then and now has been made by this technique. The technique was worked out in Germany during the early 1930s, probably initially as a way of producing a conveniently handled product. It soon became evident that the reaction was faster and produced a far superior product than reactions carried out in a single phase. The origin of this happy circumstance was the subject of extensive studies by many distinguished scientists including W. D. Harkins and P. P. Debye, who made important use of the newly developed technique of electron microscopy. It was soon recognized that the bulk of the actual reaction was occurring in small suspended particles of polymer swollen with monomer and covered with adsorbed soap. These suspended particles were all formed early in the reaction and then continued to grow in size.

The key to the whole matter was then supplied by two colleagues of mine at U.S. Rubber, W. V. Smith and R. W. Ewart,[39] who pointed out that two growing chains in the same particle rapidly would undergo mutual termination but that growing chains in different particles were isolated from each other and could not interact. As a consequence, because fresh radicals were continually being supplied to the particles through initiator decomposition, the number of growing chains at any moment was half the number of particles, and because this could easily correspond to a concentration of 10^{-7} M (much higher than radical concentrations in a typical bulk reaction), the high rate combined with the high molecular weight was explained. Roz Ewart and I shared a driving pool, and the elegance of his explanation of this long-standing puzzle one afternoon led me to drive through a red light and pay a substantial fine, a misadventure I never regretted.

The Smith–Ewart theory, which stated that the rate of an emulsion polymerization depends primarily on the number of emulsion particles present, also included a prediction of the number of particles formed as a function of the concentrations of monomers, soap, and initiator and has been validated for a number of emulsion systems. It is not well known outside of polymer chemistry, but it is interesting that there has been a recent resurgence of interest in the closely related behavior of small radicals in micellar solutions, where to some extent the same sequestering of radicals occurs. Very recently it was proposed that some of the unexpected phenomena and marked synergistic effects seen in the effect of antioxidants on the autoxidation of polyunsaturated fatty acids in heterogeneous biological systems have a similar explanation[40].

These and other developments laid much of the foundations on which both polymer chemistry and free radical chemistry have subsequently developed. I cannot avoid one ironic and disturbing fact, however. The wartime synthetic rubber program was a great success in that production rose from essentially zero in 1941 to 750,000 tons per year in 1945.[41] However, the research program and findings I have described had relatively little effect on this achievement, and the bulk of the synthetic rubber (GR-S, the emulsion copolymer of styrene–butadiene mentioned earlier) was produced by engineering modifications of a prewar German process. The greatest technical achievement of the program was probably the development of processes and plants to produce the enormous quantities of styrene and butadiene required.

Knowledge of the German process had come from pre-Pearl Harbor exchanges of information between U.S. and German companies extending back into the 1930s. This exchange included cross-licensing agreements between the Standard Oil Company of New Jersey and I. G. Farbenindustrie, in which the Standard Oil received rights for the manufacture of styrene–butadiene and acrybutrile–butadiene rubbers in the United States in return for information about an isobutylene–isoprene copolymer, "butyl rubber", which it had developed. Although this exchange has been criticized, it proved to be very much to the advantage of the United States. Butyl rubber, prepared by using a Friedel–Crafts catalyst at temperatures below −100 °C, remains a relatively high-cost specialty product, notable for chemical inertness and impermeability to gases (inner tubes for tires are a major use), whereas GR-S is still the largest volume and cheapest synthetic rubber.

During the war, speed and volume of production were crucial, and an acceptable product was already at hand. After the war, however, all the new knowledge of polymer chemistry became available to spur the postwar growth of the polymer industry. The research effort certainly was not wasted, but it took time to be applied. As an example, by 1945 it was recognized that the superior properties of natural rubber (high tensile strength and low hysteresis) were largely the consequences of its regular *cis*-1,4 structure, and it was known that the emulsion-polymerized butadiene and isoprene rubbers had irregular structures containing *cis*-1,4, *trans*-1,4 and 1,2 structures (Chart 1).

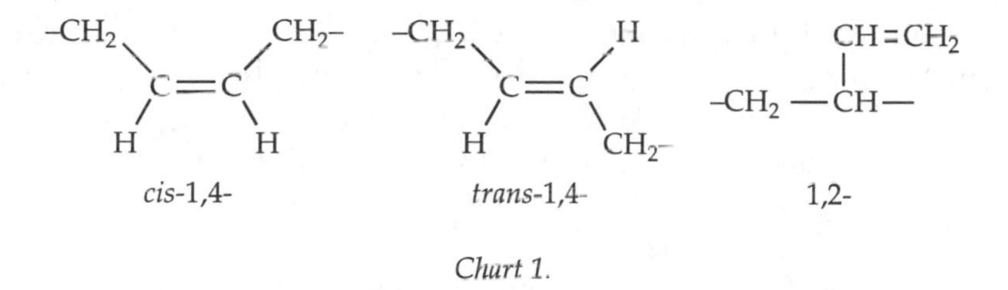

Chart 1.

However, successful development of a *cis*-1,4 polymer required two further steps: a simple, routine method of determining polymer structure, which became available with the development of commercial IR spectrometers in the late 1940s, and the discovery of new, nonradical catalysts in the 1950s, which gave the *cis*-1,4 product in high yields. As these examples show, transfer of knowledge between basic science and technology can be slow, is often unpredictable, and may flow in either direction.

Free Radical Additions

With the winding down of wartime polymer research, the interests of several groups turned back to the radical chemistry of small molecules. One such area involved radical additions yielding new carbon–carbon bonds, the fundamental step in organic synthesis.

Such a reaction might have been anticipated from Kharasch's abnormal additions, which Mayo and I had reviewed in 1940,[20] and from subsequent work on chain transfer. The 1945 report[42] from Kharasch's laboratory that, in the presence of a radical source, carbon tetrachloride and chloroform added readily to 1-octene to yield 1,1,1,3-tetrachlorononane and 1,1,1-trichlorononane, respectively, stirred up a great deal of attention, however. Actually, the same sort of reaction had been observed previously at DuPont and U.S. Rubber, and the different results obtained make an interesting example of how different equipment and sets of mind may influence the direction of research. An important wartime development had been the production of polyethylene, which proved to be a superior high-frequency insulator for radar equipment and the like. Polymerizations were run at very high pressures, and DuPont, as a producer, had investigated the use of solvents, which often led to low-molecular-weight oligomers $X(CH_2CH_2)_nY$ by chain transfer. Although this led to the names *telomer* for the oligomers and *telogen* for the species X–Y that had added, the high ethylene concentrations present at high pressures led to waxy materials with *n* being a range of relatively large numbers, products that have not had much use.

U.S. Rubber also became interested in polyethylene. A group headed by J. R. Little began a study of ethylene polymerization. Initially lacking the high-pressure equipment, they attempted the reaction in solvents at a pressure of a few atmospheres. I still recall the consternation when they reported that, in carbon tetrachloride, ethylene was rapidly taken up, but no polymer resulted. Rather, distillation yielded a mixture of low-molecular-weight liquids, $CCl_3(C_2H_4)_nCl$, with $n = 1$ to ~5! Finally, Kharasch's group, with no pressure and gas-handling equipment easily available, simply took substrates like 1-octene, which, with CCl_4, gave 1,1,1,3-tetrachlorononane and which was easily handled in ordinary laboratory glassware.

In the next few years these sorts of reactions were intensely investigated in a number of academic and industrial laboratories and they have continued to receive attention, chiefly in the USSR. A wide variety of telogens—polyhalogenated compounds, alcohols, ethers, amines, aldehydes, etc.—have been found to add to suitable olefins,[43] although, as far as I am aware, few of the processes have turned out to be commercially useful. On the other hand, as we shall see, intramolecular versions of these reactions have turned out to be powerful synthetic tools.

Autoxidation: The Faraday Society Discussion, 1947

The policy of U.S. Rubber toward its basic research groups was gratifyingly liberal, both in regard to publication (after suitable scrutiny of manuscripts by their patent lawyers) and in atten-

Frank Mayo (left) and me (right) at the U.S. Rubber Company in 1947. We had been invited to give papers at the Faraday Society Symposium, and this was a company publicity release showing us posed as though planning our trip.

dance at meetings. In the fall of 1947, Frank Mayo and I were able to attend and to present papers at a Faraday Society discussion at Oxford University. The title of the meeting was "The Labile Molecule", but it was concerned chiefly with free radical chemistry. This proved to be a most rewarding experience, because it exposed us to a large amount of British chemistry that had accumulated during the war and that was largely unpublished. The meeting also enabled us to meet a number of British chemists. What I learned had a strong influence on my thinking and turned my interests in several new directions.

Travel abroad was not the casual matter it is now. We and our wives sailed for England in considerable magnificence on the liner *Queen Mary.* Once there, besides attending the Oxford meeting, we spent several weeks traveling, visiting a number of universities and research laboratories.

At Oxford, I met W. A. Waters, whose name I had known since his 1937 review,[9] and who had just published an important and influential book.[44] Waters at the time was probably the most enthusiastic proponent of free radical chemistry in Britain. He had contributed a number of seminal ideas to the field, but I had the impression then, subsequently confirmed when I came to know him better, that he stood somewhat outside of the British organic chemical establishment, which was dominated by Sir Robert Robinson and his successors. Although Waters wrote well, he was not a forceful lecturer and his research publications were concerned more with turning up leads than reporting exhaustive studies. Nevertheless, he had a great influence on the early course of free radical chemistry and produced a loyal group of students who have carried on his work. I visited his laboratory again in 1960, and we had a last reunion in 1985 at a free radical conference at St. Andrews, at which we gave complementary talks on the early development of radical chemistry in Britain and the United States.

Other chemists I particularly recall were R. G. W. Norrish at the University of Cambridge, who was then just beginning his Nobel Prize-winning work with George Porter on flash photolysis, and H. W. Melville, then at the University of Aberdeen and already a leader in polymerization kinetics. I also had my first meeting with M. J. S. Dewar, then working with

In the Waters' garden, Oxford University, England, 1960. Left to right: W. A. Waters, Mrs. Waters, Ellie Mayo, and Frank Mayo. Waters had been trying to explain cricket to us and is now in his official robes for a university function.

C. H. Bamford at Courtaulds, Ltd., briefly in the guise of a polymer chemist.

The British view of polymerization reactions was quite different from ours, perhaps because Melville and many other polymer chemists were physical chemists by training. Our work on structure–reactivity relations was new to them. They, however, had done extensive work on polymerization kinetics and initiation reactions, including redox systems, to which I will return later. The work that had the greatest effect on my thinking was not polymer chemistry at all but the study of hydrocarbon autoxidation going on at the British Rubber Producers Research Laboratories (BRPRL) by a group including E. H. Farmer and

L. Bateman.[45] The intermediacy of hydroperoxides in autoxidations, I learned, had been well established by their work and by that of R. Criegee, H. Hock, and others in Germany, as had the peroxy radical chain mechanism of their formation with the propagation steps shown in equations 8 and 9.

$$R\cdot + O_2 \longrightarrow RO_2\cdot \tag{8}$$

$$RO_2\cdot + RH \longrightarrow ROOH + R\cdot \tag{9}$$

Oxidative degradation of rubber and other polymers was and is a serious problem, so the goal of the BRPRL work was primarily to understand the reaction so it could be avoided. However, oxygen is a cheap and ubiquitous reagent. It appeared to me that autoxidation was even more worth studying as a large-scale reaction to convert cheap hydrocarbons into more valuable materials. Accordingly, when I returned to the United States from England, I carried out an exhaustive survey of the literature on autoxidation and peroxide chemistry, and we tried hard to interest our management in a comprehensive research program. Some work was started, but with little management support, the project never really flew. Although my assessment of the promise was correct (at present, some 13 of the 30 largest volume organic chemicals are made by oxidation reactions involving molecular oxygen), the subsequent successful applications of autoxidation to synthesis occurred elsewhere.

Lever Brothers Company, 1949–1952

By mid-1948, research in free radical chemistry and my enthusiasm for it were booming, but the situation at U.S. Rubber was not. Five years of work had not led to major new products, there were important changes in management, and it became evident that support of basic research for its own sake would not continue for long. What was going to happen instead was not at all clear, and in the fall of 1949, in the absence of an attractive academic opportunity, I accepted a position as head of organic research at Lever Brothers Company in Cambridge, Massachusetts.

My move was part of a rather extensive exodus. Among others, E. J. Hart and M. S. Matheson moved to the Argonne National Laboratories to work in radiation chemistry, and Fred Lewis, and then Frank Mayo, moved to the General Electric Company. Although U.S. Rubber's general laboratories themselves were soon moved to much more elegant facilities in Wayne, New Jersey, the sort of research we had been doing came to an end.

At Lever Brothers Company I was largely concerned with synthetic detergents and the development of new products and processes. It was a good chance to apply my ideas of physical organic chemistry to practical synthesis and also to learn some of the arcane oddities of a consumer-oriented industry.

An example of how unanticipated criteria can frustrate industrial research was a project to develop an odorless permanent wave for home use. The active ingredient of most preparations was some salt of thioglycollic acid in a weakly alkaline medium. It acted by reducing some of the disulfide linkages in hair protein, making it plastic. After setting in a wave, the hair was then fixed by reoxidization. Thioglycollic acid has a pronounced odor, so we developed a practical synthesis for β-mercaptoethanesulfonic acid, which was nonvolatile and odorless. Its salts were effective, and we carried the process as far as a semiworks including a large-scale ion-exchange step to give salt-free solutions of the acid. The project failed on two counts: reduced hair itself has a significant odor, and the company economists decided that there was too little money in the business.

I was also in a good position to keep in touch with the rapid development of physical organic chemistry. J. D. Roberts, C. G. Swain, and W. H. Stockmayer were all just down the street at MIT, and I saw them frequently. Seminars and meetings at both MIT and Harvard were available, and I once had the temerity to address the joint Harvard–MIT physical chemistry seminar on some experiments I had done[46] involving application of the Hammett acidity function principle to measurement of the acidities of the surfaces of silica–alumina and similar catalysts by observing the colors of adsorbed indicators. This was one of the

Participants at the 3rd Reaction Mechanisms Conference at Northwestern University in 1950. Highlighted participants from left to right: C. Wilson, me, Saul Winstein, Jack Roberts, C. K. Ingold, A. A. Morton, unknown, P. A. S. Smith, M. Newman, R. Pearson, R. H. Baker, D. P. Stevenson, R. Letzinger, and H. C. Brown.

Christopher K. Ingold (at right) at the 3rd Reaction Mechanisms Conference, Northwestern University, 1950. This was his first visit to the U.S.

first indications that such materials, under anhydrous conditions, could be extremely strong acids.

I also saw a good deal of Paul Bartlett at Harvard; I attended his group seminars regularly and even went on weekend bicycle trips with him and our older children. The seminars were largely devoted to reviews of the work of his students, then divided between studies of solvolyses and investigations of peroxide chemistry, but we also had visiting speakers. I recall Jack Roberts spending several sessions trying to convince us that molecular orbital theory was both comprehensible and useful. Jack has always been a missionary for new ideas and, a few years later, did much to sell NMR spectroscopy to organic chemists.

My management was also generous in sending me to meetings further afield, and in 1950 I served as chairman of the Third Conference on Organic Reaction Mechanisms held at Northwestern University. There I reviewed the data on structure and reactivities deduced from copolymerization and attempted to apply our conclusions, notably the importance of the polar effect, to the then-available data on other radical reactions. To most participants, however, the most notable feature of the meeting probably was the presence of C. K. Ingold, the dean of British physical organic chemists, making his first visit to the United States. Ingold gave a distinguished lecture and charmed all of us who had never met him previously.

Louis Plack Hammett, October 1954.

A Goal Realized—Academic Research Columbia University, 1952–1969

My long-term goal, however, remained to return to basic research, preferably at a university. The summer of 1952 became something of a deadline, because our laboratories were due to move to Edgewater, New Jersey, at that time. My feelers to the academic community were not receiving much of a response; however, in that spring, I was approached by Louis Hammett, then the chairman at Columbia University, where a professorship might be open. I was a great admirer of Louis, whom I had come to know quite well through my earlier work on the application of the Hammett equation. We had discussed academic positions several times in the past, and I responded with enthusiasm. The only obstacle was that the position had been offered to Jack Roberts. Fortunately for me, Roberts and his wife both preferred the West Coast. He moved to Caltech (California Institute of Technology), and I received the appointment. Besides my debt to Louis for his faith in me, I have always felt a debt to Jack for his geographic preferences.

High-Pressure Reactions

My appointment at Columbia left me, for the first time, free to develop my own research program, a matter that I approached with some trepidation. One project I had picked up in my reading was the examination of the effect of pressure on the course and rates of liquid-phase reactions and the determination of transition-state volumes via the relation in equation 10 as an additional tool in investigating reaction mechanism.

$$\frac{-d \ln k}{dP} = \frac{\Delta V^{\pm}}{RT} \tag{10}$$

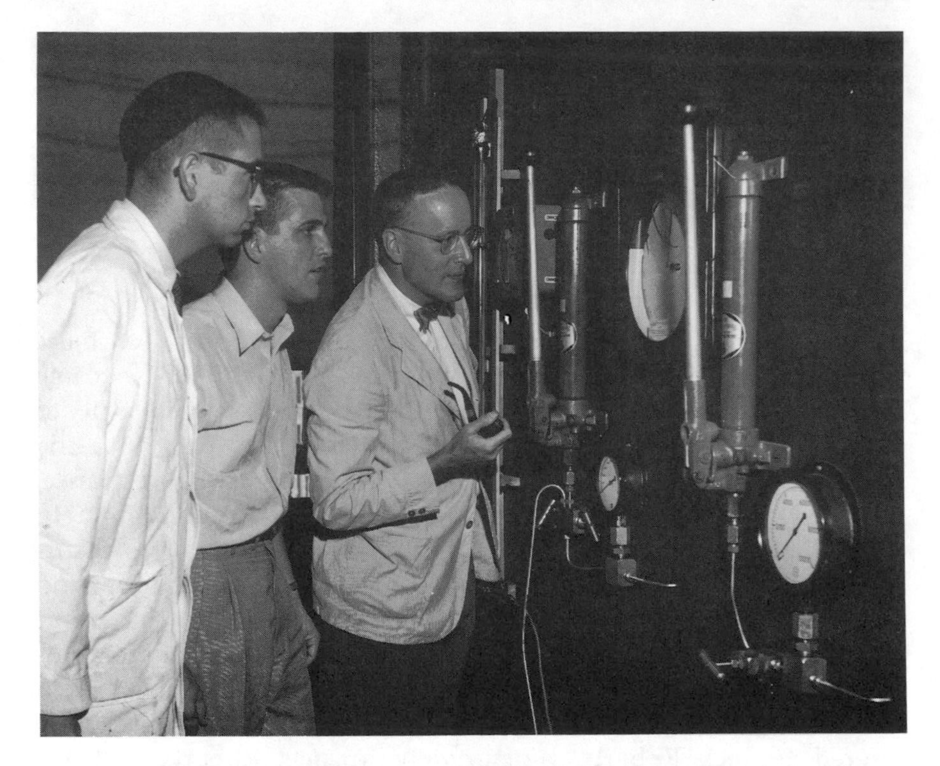

Our high-pressure laboratory at Columbia University, 1955. Left to right: Jack Peisach, Joseph Pellon, and myself. The view shows the control panel and the hand-driven pumps with which Jack and Joe had to struggle. The actual equipment is behind the steel barricade. Jack was studying Diels–Alder reactions while Joe did our work on polymerization.

In equation 10, $\Delta V^{\pm}$ represents the volume change in going from reactants to transition state. A negative value (decrease) predicts an increase in rate with pressure. A limited amount of work had been published in the area, and it was evident that, in order to produce significant changes in rates, pressures in the range of 1–10 kbar were required.

Fortunately, through my uncle, Reginald A. Daly, emeritus professor of geology at Harvard, I had an introduction to P. W. Bridgeman in the Harvard physics department. Bridgeman was a Nobel prize winner and, at that time, the grand old man of high-pressure technology. While I was still at Cambridge, I was able to visit him to see how high-pressure experiments were actually done and to get advice on where to obtain the necessary equipment. He received me cordially and, as a demonstration, kindly attempted (unsuccessfully) to effect for me the Diels–Alder reaction between furan and dimethylmaleic anhydride at 20 kbar as a possible short-cut synthesis of cantharidin. I had just been introduced to this molecule in a seminar by Gilbert Stork, who was also about to join the Columbia faculty.

Our own work at Columbia in this area began in 1953–1954, when I was able to order suitable equipment and to recruit two graduate students, Joseph Pellon and Jack Peisach, to the cause. Following Bridgeman's technique, we produced the necessary pressures by using an "intensifier" (a large piston driving a smaller one), with the large piston driven, originally, by a hand-operated hydraulic pump. With it only my strongest students were good to 12 kbar, the limit of the equipment.

Our high-pressure work continued into the 1960s and led to a number of publications and theses. Study of Diels–Alder reactions[47,48] and other electrocyclic processes[49] showed that all are strongly pressure accelerated, a fact implying small transition states. The polymerization of butyraldehyde to a high-molecular-weight polyacetal was shown to be a simple acid- and base-catalyzed, reversible process, which occurred readily at high pressures because of the volume decrease but had a ceiling temperature well below 0 °C at normal atmospheric pressure.[50] Finally, we were able to determine the effect of pressure on the individual steps of chain propagation and chain transfer in some radical polymerizations and on peroxide decompositions.[51,52]

The polymerization work took a little ingenuity. It was known that the overall rates of radical polymerizations were generally pressure accelerated, but it was unclear which steps were speeded up. However the Smith–Ewart theory of emulsion polymerization, with which I was familiar from my U.S. Rubber days, stated that, once an emulsion polymerization was started, the rate depended only on the concentration of emulsion particles and the rate constant for chain propagation. Accordingly, our approach was to start an emulsion polymerization (conveniently of styrene), divide it into two portions, put one in the pressure equipment, and then compare the two rates. At 5 kbar, the rate increased about 10-fold, corresponding to $\Delta V^{\pm} = -11.5$ mL/mol. With this number in hand, measurements of the effect of pressure on the molecular weight of polymers produced in

With M. G. Gonikberg (right) at a polymer meeting in Moscow, 1960. I gave a paper on our work on the effect of high pressure on vinyl polymerization, and Gonikberg had been a pioneer in the study of the effect of pressure on reaction rates.

the presence of chain-transfer agents like CCl_4 showed negligible effects, results indicating a similar $\Delta V^{\pm}$ for the chain-transfer step. Taking all data together, we concluded that, in an ordinary, peroxide-initiated polymerization, the overall acceleration due to pressure is due to an increase in the rate of chain propagation, together with compensating decreases in the rates of chain initiation and chain termination.

I presented this polymer work at a symposium on polymer chemistry in Moscow in 1960 as part of a European tour that Frank Mayo, our wives, and I took together in that year. Because 1960 was the year of the "U-2 incident," we were somewhat apprehensive of our reception in the USSR. Everyone, however, was cordial, and the entertainment was lavish. I found the USSR more colorful than I had anticipated, but that research was spotty and handicapped by a lack of equipment. Some work was highly original, but much was rather routine, with institutes run rather like industrial laboratories at home.

Although I terminated our pressure work in the 1960s, the use of pressure to determine transition-state volumes has continued in other laboratories, including that of R. C. Neuman at the University of California at Riverside, who began his work in this area with me. W. G. Dauben has recently demonstrated the synthetic utility of high pressures in carrying out difficult and highly hindered Diels–Alder reactions so that it continues to be a useful technique.

Free Radicals in Solution

My second project involved writing. I had learned a great deal and enjoyed preparation of the reviews I had written with Frank Mayo, and at the time I went to Columbia, I was writing two chapters on vinyl polymerization and Diels–Alder reactions for a book on the chemistry of petroleum hydrocarbons.[53] I now began to consider a more ambitious project, a comprehensive monograph on free radical reactions in the liquid phase, a topic not thoroughly covered since Water's book of 1946.[44] In anticipation, I kept up the survey of peroxide chemistry I had begun at the U.S. Rubber Company and began to collect references for other reactions as well. Actual writing began in the spring of 1954; the manuscript was finished in 1956; and the result, *Free Radicals in Solution,*[54] appeared in 1957. I found the time for this book by doing most of the writing during my summers and by taking advantage of the fact that Columbia, until the late 1960s, provided no parking facilities for its staff. This gave me ample time for my reading on the bus and subway while commuting from New Jersey. Finally I got double duty from the developing manuscript by using it as the basis for a graduate course on radical chemistry.

The book was well received and, I like to think, had considerable impact in spreading interest in and understanding of free radical chemistry. It was used as a graduate textbook, had extensive sales to industrial laboratories, and was shortly translated into Russian. Much of its success was certainly due to fortunate timing. It filled a gap and came out at a time when the field was small enough that almost all significant references could be included. A few years later, when I might have considered a second edition, I realized that so much had been published that any attempt to cover all of free radical chemistry in a single book by one author could produce no more than an introduction. Although it has long been out of print it remains a good guide to older literature and shows how much has been rediscovered since.

New Lines of Free Radical Research

My third project, the largest and most important, was to resume my research in free radical chemistry. The question was where to start. I adopted the scheme I have used ever since—to keep a file of problems that seemed significant and for which I had ideas for at least an initial experimental approach and then try to interest potential students or other collaborators in actually undertaking some problem about which I was particularly excited at the time. This approach lacks the elegance of a single-minded attack on a unique problem and has meant that my research has always followed a rather zigzag and unpredictable course.

The first two problems came straight from my file on autoxidations and peroxide chemistry. One involved the reported observation that *t*-butyl hydroperoxide, although thermally very stable, was an effective polymerization initiator at relatively low temperatures. Initial work was done by a postdoctoral fellow, Y.-W. Chang,[55] with much further detail and understanding provided later by LaDonne Heaton in her thesis:[56] With monomers like styrene, hydroperoxides undergo a rapid bimolecular reaction, but styrene oxide is the major product, with only a small yield of chain-initiating radicals. This dichotomy between peroxide decompositions yielding radical and nonradical products was one to which I returned later and will discuss later in this account.

The second problem, undertaken with my first graduate student, Sheldon Buckler, was to examine the autoxidation of Grignard reagents. Porter and Steele, in 1920, had proposed a two-step sequence to the known alcohol products (equations 11 and 12).

$$\mathrm{RMgX} + \mathrm{O_2} \longrightarrow \mathrm{ROOMgX} \qquad (11)$$

$$\mathrm{ROOMgX} + \mathrm{RMgX} \longrightarrow \mathrm{2ROMgX} \qquad (12)$$

Such reactions had always been run by bubbling dry air through Grignard solutions. By reversing the procedure and slowly adding the RMgX solution to cold, O_2-saturated ether, Buckler[57] was able to obtain high yields of the intermediate peroxide and to validate the reaction scheme. With some extensions, this work has provided a useful synthesis for a variety of hydroperoxides. Our efforts to establish the mechanism of the initial autoxidation (which is now clearly a radical chain) were less conclusive, and all we could show was that the reaction was extremely rapid. Substituting acetaldehyde for ether as the O_2-saturated solvent gave no aldehyde consumption and similar peroxide yields.

A third early problem, taken on by my second graduate student, Bernard Miller, arose from my interest in structure–reactivity relations and the Hammett equation. This was to examine the effect of ring substitution on the radical side-chain chlorination of toluene. The results we had anticipated was that Cl· should be an electrophilic radical, and as we hoped, the reaction showed a negative ρ value (–0.78).[58] However, the analytical technique—competition experiments with α,α,α-trideuteriotoluene, conversion of the DCl–HCl mixture to H_2O–D_2O, and analysis by density with a density gradient column—was distressingly tedious.

Miller's enthusiasm and perseverance survived, and he even took on a related problem, the photoinduced reaction of

Cl_2 with aryl bromides to yield aryl chlorides and Br_2. This was a reaction I had inadvertently stumbled on at U.S. Rubber in the course of a routine synthesis, only to find it had been reported as early as 1903 by Eibner[59] and periodically rediscovered ever since. The reaction turned out to be remarkably facile, occurring 1/4 to 1/5 as rapidly as side-chain chlorination of toluene in competitive reactions.[60] This result and a small solvent effect noted in the toluene chlorination led me to suspect that significant solvent effects might be observed in chlorinations carried out in aromatic solvents. In this regard, I was anticipated by G. A. Russell,[61] who, in 1957, showed that aromatic solvents indeed greatly increased selectivity in alkane chlorinations and ascribed the effect to π complexation of the chlorine atoms. Russell's paper came out as we were waiting for starting materials, and we subsequently confirmed his observations and concluded that the higher selectivities of the complexed species resulted from slower reactions of higher activation energies.[62] This, in turn, has more recently been confirmed by K. U. Ingold et al.,[63] who have actually measured the rates of the individual steps by laser flash photolysis.

Left to right: Glen Russell, Peter Wagner, and Bill Pryor. Three free radical chemists helping me celebrate my ACS award in petroleum chemistry at the spring ACS meeting in St. Louis, 1984.

Hypochlorite Chemistry and Halogen Carriers

An item on my "idea list" that I brought back from discussions at the 1952 reaction mechanisms conference at Bryn Mawr involved the use of *t*-butyl hypochlorite as a chlorinating agent, for which scattered evidence existed in the literature. At that time *N*-bromosuccinimide (NBS) was generally believed to react by the Bloomfield mechanism,[64] involving the succinimide radical as a chain carrier (equations 13 and 14).

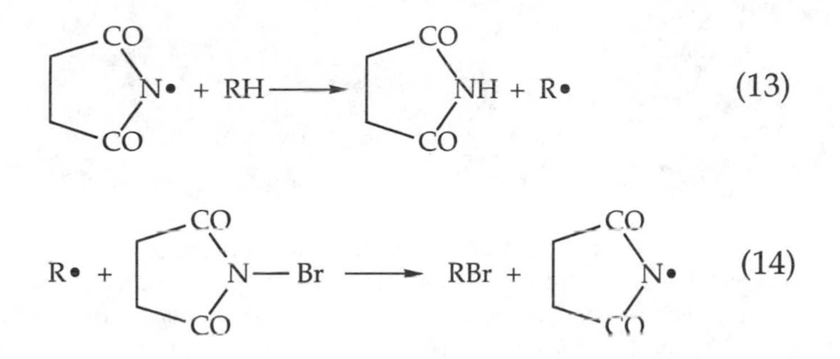

I therefore postulated a similar sequence for the hypochlorite involving the *t*-butoxy radical (equations 15 and 16).

$$(CH_3)_3CO\cdot + RH \longrightarrow (CH_3)_3COH + R\cdot \qquad (15)$$

$$R\cdot + (CH_3)_3COCl \longrightarrow RCl + (CH_3)_3CO\cdot \qquad (16)$$

A student, B. B. Jacknow, began work on the problem in 1954. He demonstrated that the reaction competed with the known β scission of *t*-butoxy radicals (equation 17) and that *t*-butoxy radicals showed quite a different selectivity from that of chlorine atoms.[65]

$$(CH_3)_3CO\cdot \longrightarrow CH_3COCH_3 + CH_3\cdot \qquad (17)$$

This work led my group into a very fruitful area of research that extended for almost 20 years and about which I can only summarize the high points.

At Park City, Utah, 1986, twenty years after the last photo. Left to right: Phil Rieger (Brown University), Nancy Rieger, Al Padwa (Emory University), Pete Wagner (Michigan State), and Mike Mintz (Dow). All were graduate students at Columbia in the 60s and (with the exception of Phil, who worked with George Fraenkel) were members of my research group. The occasion is a symposium marking my 70th birthday.

First, the chemistry of alkoxy radicals was of interest because they are products of peroxide decompositions and may be produced in a variety of other ways. The facile chlorination of a variety of substrates not only showed that *t*-butyl hypochlorite could be a useful chlorinating agent but also yielded extensive data on the relative reactivities of C–H bonds toward alkoxy radicals. Because the bond dissociation energy of *t*-butyl alcohol is ~104 kcal, the energetics of equation 15 is very similar to that of Cl· reactions. However, reactions are slower and selectivities are considerably higher, with the relative reactivities of 1°, 2°, and 3° C–H bonds being approximately 1:12:44. Alkoxy radicals are also strongly electrophilic species. They selectively attack the α-C–H bonds of ethers and the RCO–H bonds of aldehydes. They are, however, sufficiently unreactive toward acetic acid and acetonitrile that these solvents may be used for hydrocarbon chlorinations.

Second, the hypochlorite proved to be a very selective agent for chlorination at allylic positions, with little tendency to add to double bonds. Cyclohexene is converted to 3-chlorocyclohexene in high yield, and its allylic C–H bonds are some 15 times as reactive as the C–H bonds of cyclohexane. Further, because allylic chlorides are relatively stable toward isomerization, we could show, for the first time, that *cis* and *trans* allylic radicals could retain their configurational integrity.[66]

Thus, *cis*- and *trans*-2-butene yielded different products (Scheme 2). In more complex molecules, products formed without a shift of the double bond retain their configuration, but those in which the bond has shifted are formed chiefly in the more stable *trans* configuration.

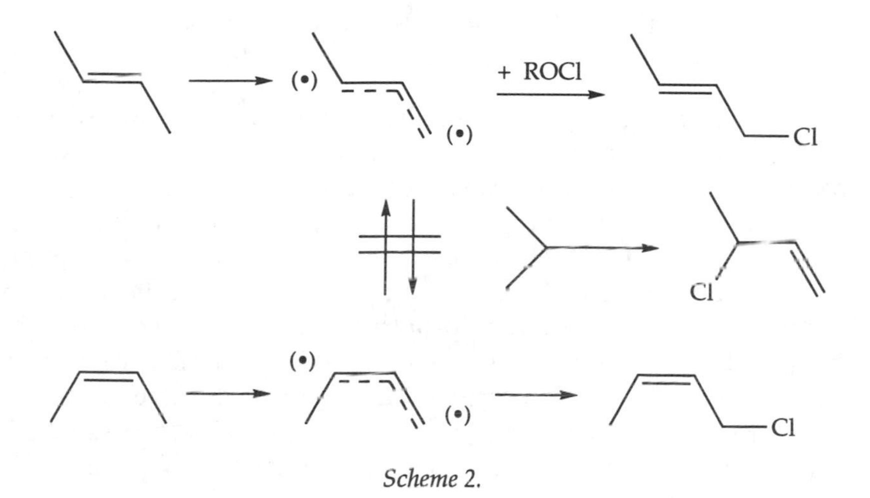

Scheme 2.

Next, Peter Wagner,[67] in a very careful study, showed that solvent effects were significant in the competition between hydrogen abstraction and β scission but not in C–H bond selectivity. His conclusion was that, because solvation chiefly involves the alkoxy oxygen, the transition state for β scission could be solvated but not that for C–H bond attack.

Finally, Albert Padwa[68] examined the β-scission pattern of alkoxy radicals derived from hypochlorites with other larger side chains and showed that, with long side chains, intramolecular chlorination becomes the dominant reaction, occurring preferentially through quasi-six-membered transition states to give δ-chloroalcohols. Such intramolecular reactions occur with alkoxy radicals from other sources, as in the photolysis of alkyl nitrites (the Barton reaction) and the oxidation of alcohols with lead tetraacetate, and have come to have extensive synthetic use. They also provide a diagnostic test for the presence of alkoxy radicals in oxidative reactions.

I noted that our hypochlorite work was originally suggested by analogy to the then-current view of NBS chemistry, and the two reactions have remained oddly intertwined. An alternate formulation of NBS reactions is the Goldfinger mechanism,[69] in which bromine atoms are the chain carrier, and NBS simply serves as the Br_2 source via a fast reaction (equation 18).

$$HBr + NBS \longrightarrow Br_2 + \text{succinimide} \tag{18}$$

Papers from my group and elsewhere in 1963[70–72] supported the Goldfinger mechanism on the basis of selectivity patterns, and the Bloomfield mechanism went into eclipse.

Next, some puzzling results in hypochlorite chlorinations were shown by Jim McGuinness[73] to be due to a similar incursion of chlorine atom chains there but that these could be readily eliminated by adding dichloroethylene or a similar olefin as a Cl• trap. This simple answer was not arrived at easily. As our paper stated, "The order in which our arguments. . . are presented. . . has little relation to the chronological sequence in which apparently unrelated or contradictory data were accumulated." Starting with some false assumptions, we pursued a number of false leads, and I can only admire Jim's patience and persistence along the way. This technique of using traps made it feasible to study primary and secondary hypochlorites, where Cl• chains are much more important,[74,75] and enabled Skell,[76] a bit later, to suppress Br• chains in NBS reactions. The Bloomfield mechanism reemerged, and it now appears that NBS and its analogues can react by either mechanism. In allylic and benzylic brominations, where these reagents are usually employed, Br• chains predominate, but with less reactive substrates in the presence of Br• traps, succinimide radicals compete. In such reactions, halogenation is accompanied by extensive ring opening of the succinimide radical to give *β*-bromopropionyl isocyanate[77,78] (equation 19).

$$\text{succinimidyl radical } (\mathrm{N}\cdot) \longrightarrow \mathrm{OCN{-}CO{-}CH_2\dot{C}H_2} \xrightarrow{\mathrm{NBS}} \mathrm{OCN{-}CO{-}CH_2CH_2Br} \quad (19)$$

Thiyl and Phosphoranyl Radicals

The use of thiols as transfer agents to control molecular weight in polymerization had attracted my attention to their high reactivity in radical reactions while I was at U.S. Rubber. The chain steps in thiol additions are shown in equations 20 and 21.

$$\mathrm{RS\cdot + CH_2{=}CHR'} \underset{k_{-a}}{\overset{k_a}{\rightleftarrows}} \mathrm{RSCH_2\dot{C}HR'} \quad (20)$$

$$\mathrm{RSCH_2\dot{C}HR' + RSH} \underset{k_{-d}}{\overset{k_d}{\rightleftarrows}} \mathrm{RSCH_2CH_2R' + RS\cdot} \quad (21)$$

Because there was evidence that the addition step was reversible, Wolf Helmreich began a study of this reversibility and its effect on the apparent relative reactivities of different olefins in competitive experiments. The work came out nicely. For the addition of methanethiol to *cis*-2-butene, $k_{-a} \geq 80\ k_d$, leading to extensive isomerization of unreacted olefin, and the relative reactivities of a number of olefins were determined.[79] It was also a landmark because of our first use of gas–liquid chromatography (GLC) for product analysis. This technique had just become available and was ideally suited for our work. Analyses of mixtures, which have previously required ingenious techniques, large samples, and days to weeks of effort, could now be done on a few drops of reaction mixtures in minutes. I became an enthusiastic promoter of the technique. In lectures at the time, I probably gave it as much attention as I gave interpretation of our results.

At about the same time, another student, Bob Rabinowitz, examined the reversibility of the displacement step, that is the ease of thiyl radical attack on C–H bonds. The technique involved the photolysis of isobutyl disulfide in the presence of hydrocarbon substrates and following the production of thiol. Cumene, for example, was easily attacked. We arrived, however, at the interesting conclusion that, in aromatic solvents, the light was chiefly absorbed by the solvent, which then sensitized the disulfide dissociation.[80] Photochemistry had not yet taken the prominent place in physical organic chemistry that it now occupies, and this was my first exposure to this sort of energy transfer, of which more later.

In reading up on thiol chemistry, Bob came across a report by F. W. Hoffman[81] that attempts to carry out the exchange reaction between a thiol and trialkyl phosphite led, instead, to the remarkable oxidation–reduction reaction in equation 22.

$$\mathrm{RSH + P(OR')_3 \longrightarrow RH + SP(OR')_3} \qquad (22)$$

The reaction was reported to be light catalyzed, and Bob proposed a chain sequence (equations 23–25) involving an intermediate phosphoranyl radical ($R_4P\cdot$), a species that had just been proposed and named by Fausto Ramirez,[82] then a colleague at Columbia.

$$\mathrm{RS\cdot + P(OR)_3 \longrightarrow RS\dot{P}(OR')_3} \qquad (23)$$

$$\mathrm{RS\dot{P}(OR')_3 \longrightarrow SP(OR')_3 + R\cdot} \qquad (24)$$

$$\mathrm{R\cdot + RSH \longrightarrow RH + RS\cdot} \qquad (25)$$

Bob quickly established the radical chain nature of the reaction, showed that it also occurred with disulfides, and demonstrated a similar nonchain process with alkoxy radicals.[83] These results launched us, for a time, into studies of the chemistry of phosphoranyl radicals.

Radical Cyclizations

The thiol–phosphite reaction just described provided a way of generating a radical center at a specific point in a complex molecule as a step in a chain reaction. It occurred to me that it could be used to study radical cyclizations, an area that, at that point, had received very little attention. Myrna Pearson examined the reaction of 6-mercapto-1-hexene with triethyl phosphite. We had expected the cyclization shown in equation 26, but we were astonished to obtain almost entirely methylcyclopentane.[84]

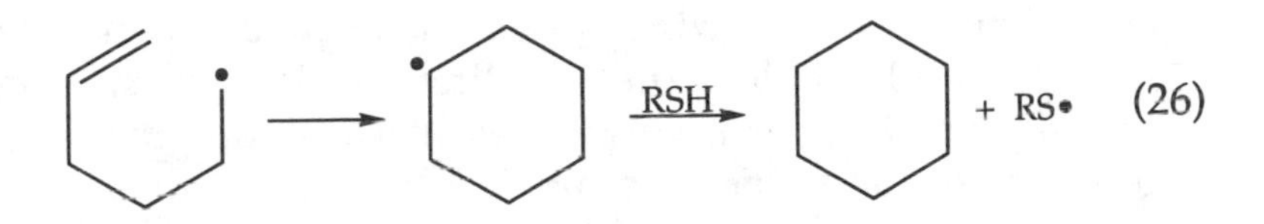

By the time our work was published, other examples of such closures of 5-hexenyl- to cyclopentylmethyl radicals had appeared,

Marc Julia, Mme. Julia, and Jane in Park City, Utah, 1980. Marc was one of the first investigators of free radical cyclization reactions. I met him first at the Gomberg Symposium at Michigan in 1966 and he was subsequently my host in Paris. He lectured at the University of Utah in 1980.

and the factors determining 5- versus 6-membered-ring closure have now been extensively studied.[85]

Unsaturated thiols readily polymerize or cyclize to thioethers. Because yields of carbocyclic products in their reaction with phosphites are rather low, we did not pursue this particular cyclization scheme further. I kept my eyes open, however, for other ways of generating radicals regiospecifically for cyclization studies. A lead came in the radical chain reaction between alkyl halides and tin hydrides demonstrated by Menapace and Kuivila.[86] J. H. Cooley, then visiting my laboratory from the University of Idaho and working with two undergraduates, was able to show that several 6-bromo-1-hexene derivatives could be cyclized readily by tributyltin hydride[87] to give chiefly cyclopentane derivatives via the sequence shown in Scheme 3.

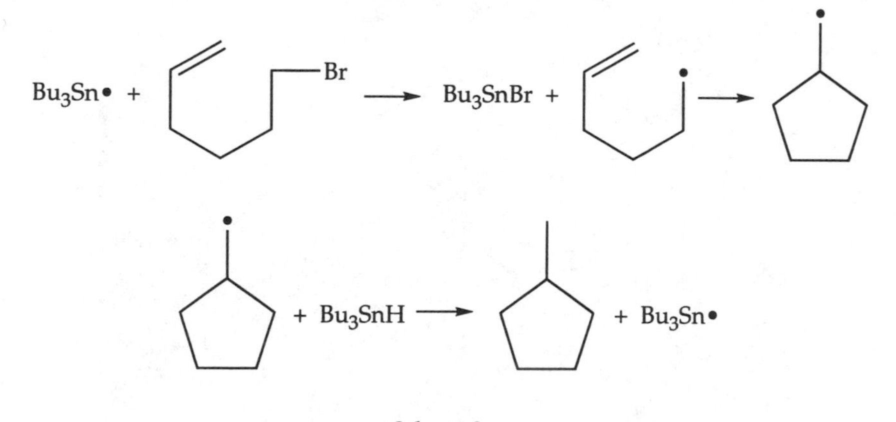

Scheme 3.

Angela Cioffari,[88] using the reaction, undertook a more detailed study of the regio- and stereochemistry of cyclization of a number of substituted 6-bromo-1-hexenes and also the relative rates of cyclization and reduction of the intermediate 5-hexenyl radicals by the hydride that determined cyclization yields. Her work further demonstrated the scope and utility of the reaction, which has subsequently been widely used and has become, with extensions, a method of choice in the growing application of such cyclizations to organic synthesis. In retrospect, it has proven to be the most synthetically useful reaction with which I have been involved.

Peroxide Chemistry

Over the duration of my chemical career, the chemistry of organic peroxides has developed into a real jungle of fact and theory, involving phenomena from biology to chemiluminescence. My research group and I have made several forays into this underbrush, and I shall discuss three problems we explored.

There is a class of reactions in which peroxides react with a variety of electron-rich substrates, sometimes to give radicals in significant yields, but often not. Two limiting mechanisms are possible: simple nucleophilic attack on the peroxide by an electron pair of the substrate or single electron transfer (SET) to yield radical ions. These alternative mechanisms have proved difficult to distinguish, and an intermediate "merged mechanism," with electron transfer contributing to a partially bonded transition state (analogous to the participation of polar structures in a radical transition state) is also conceivable.

In the 1950s, these ideas were just emerging, and we examined the reactions of benzoyl peroxides and similar peroxides with tertiary amines and phenols. The amine reactions were known to be practical, low-temperature radical sources, but I don't think we contributed much except to show that they were

Chengxue Zhao and myself in my office at Utah surrounded by a bevy of secretaries. He wanted a picture of them all to take back to China.

complex and the radical yields were low.[89] The phenol reactions were more tractable. We found no radical production and simple relations with phenol structure: acceleration by electron-supplying groups, and steric retardation on 2,6 substitution. We interpreted both reactions as nucleophilic processes.[90]

Twenty years later, Chengxue Zhao, a visiting scholar from the Shanghai Institute of Organic Chemistry, and I returned to the area, which had developed a large literature. We used several ring-substituted benzoyl peroxides and a series of dimethoxybenzenes as substrates.[91] These showed reactivities comparable to those of the phenols, with rates paralleling the oxidation potentials of the substrates and the reduction potentials of the peroxides. The products were quite varied, including both ring- and (with methyl-substituted dimethoxybenzenes) side-chain-substituted products. We concluded that we were dealing with an SET reaction, yielding an initial radical ion pair (Scheme 4) but one involving considerable substrate–peroxide bonding (i.e., a "merged" mechanism), because there was also evidence for marked steric hindrance.

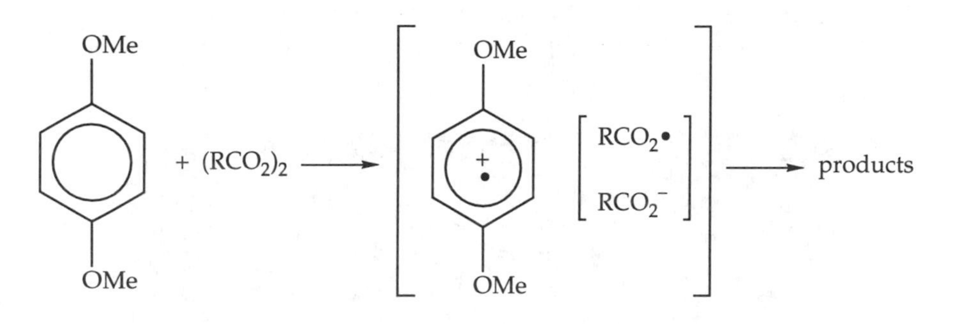

Scheme 4.

With suitably negatively substituted peroxides, these reactions are remarkably facile. Zhao brought with him from China an interest in perfluoroacyl peroxides and was able to show that perfluorobutyryl peroxide reacted rapidly with dimethoxybenzenes, even at 0 °C.[92]

Another rapid peroxide decomposition that was extensively studied in the 1950s and the 1960s involved molecules such as phenylacetyl peroxide,[93] believed to decompose by the concerted scission of two or more bonds to yield relatively stable radicals (e.g., with phenylacetyl peroxide, the benzyl radical; equation 27).

$$\text{RC(=O)OOC(=O)R} \longrightarrow \text{R}\bullet + CO_2 + \bullet\text{OCOR} \quad (27)$$

The rates of these decompositions proved to be quite dependent on solvent polarity, and the yields of radicals were variable and often quite low. Further study showed that the major, initial nonradical product was an acyl carbonic anhydride (the so-called carboxyl inversion product) arising from a typical cationic rearrangement, which subsequently decomposed largely to an ester (equation 28).

$$\text{RC(=O)OOC(=O)R} \longrightarrow \text{R—O—CO-O-COR} \longrightarrow \text{ROCOR} + CO_2 \quad (28)$$

I became interested in these apparent competing radical and polar paths of decomposition, and so we undertook a detailed study of two typical cases: isobutyryl peroxide and isobutyryl *m*-chlorobenzoyl peroxide.[94] We found that ratios of radical products to polar products were quite dependent on peroxide structure but less so on solvent. This and other considerations led us to propose that all products arise from a single rate-determining transition state, leading to a radical pair–ion pair that subsequently partitions to radical and ionic products (Scheme 5).

In Scheme 5, N represents a nucleophilic solvent, and the radical pair–ion pair may be thought of as a resonance hybrid or as two species in equilibrium depending on their separation. Again, as with the concepts of polar effects and merged mechanisms, we are on the borderline between radical and ionic processes. Although this common transition-state model has generally been accepted as plausible, it is difficult to prove. Our subsequent work has been aimed at investigating the timing of

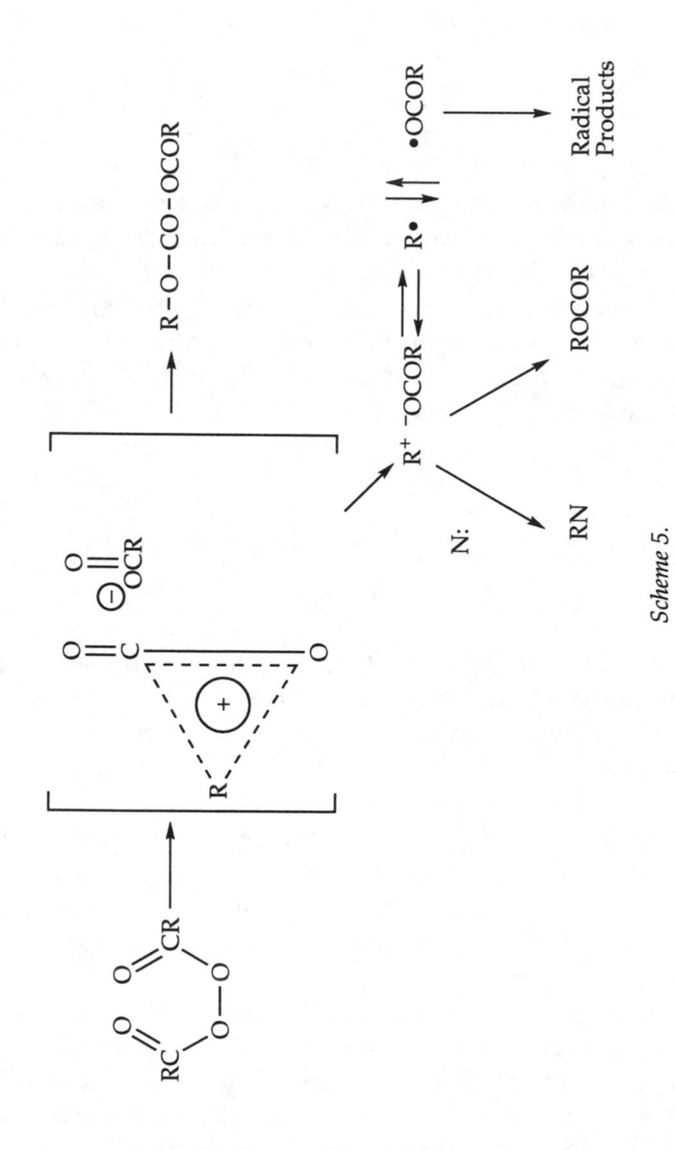

Scheme 5.

the various steps by using stereochemical probes. Thus, with peroxide derived from optically active 2-methylbutanoic acid, carboxyl inversion occurs early with complete or almost complete retention of configuration, but subsequent reaction of the ion pair with nucleophilic solvents, for example, acetonitrile or acetic acid, shows moderate net inversion, similar to that observed in classical solvolyses.[95,96] Taylor[97] has reported similar results, but probably the best evidence for radical–ion pair interconversion has been the detection of chemically induced dynamic nuclear polarization (CIDNP) in the NMR spectra of the products of ionic rearrangements in some systems observed by Lawler.[98]

Although the susceptibility of organic peroxides to photolytic dissociation had long been known, in 1955 I had noted a report by Luner and Szwarc[99] that the decomposition of acetyl peroxide could be photosensitized by anthracene and pointed out in my book[54] that such photosensitizations were interesting and might have practical use. A few years later, I was able to interest a graduate student, Morton J. Gibian, in the problem. He showed that both aromatic compounds and ketones were efficient photosensitizers for a variety of peroxides and that the reaction with the ketones, at least, involved energy transfer from the triplet state.[100]

The rich chemistry of triplet-state ketones was receiving much attention at this time. Gibian went on to show that their ability to attack R–H bonds paralleled closely the behavior of alkoxy radicals.[101] Such sensitized photolyses of peroxides have subsequently been the subject of considerable study, but the most striking and useful observation in our work involved photolysis of diacyl peroxides of the sort discussed in the previous section, which decompose thermally by multibond scission to give a mixture of ionic and radical products.

Triplet energy transfer and dissociation, we reasoned, should lead to triplet radical pairs that could not go to ion pairs without prior spin inversion and so could not easily give rise to ionic products. This was demonstrated for the peroxide from *trans*-4-*t*-butylcyclohexane carboxylic acid, which, thermally, gives ~90% ionic products, chiefly the carboxyl inversion product. In contrast, benzophenone-sensitized photolysis in CCl_4 gave 66% C_2Cl_6, plus 4-*t*-butylcyclohexyl chloride, clearly derived from radical intermediates. This dichotomy has proved to be quite general and is useful in providing a way of generating specific radicals in good yield from the corresponding diacyl peroxides, even in cases where thermal decompositions lead to nonradical products.

Actually, as it turned out, we had had an earlier clue to this dichotomy. A few years before, A. Naglieri,[102] investigating a series of nitrogen analogues of benzoyl peroxide, had found that *O,N*-dibenzoylhydroxylamine decomposed thermally to phenyl isocyanate and benzoic acid, with only a trace of radical production (equation 29).

$$C_6H_5CO{-}NH{-}OCOC_6H_5 \longrightarrow C_6H_5NCO + C_6H_5COOH \quad (29)$$

The reaction is an analogue of the carboxyl inversion reaction. In contrast, photolysis yielded benzene, benzoic acid, and benzamide, and both *O,N*-dibenzoylhydroxylamine and dibenzoylhydrazine were good photosensitizers for radical polymerizations.

The Academic Scene at Columbia

My arrival at Columbia in the summer of 1952 dropped me into the middle of an academic world with which I had had only indirect experience. I was appreciative of much valuable advice I received, both from Louis Hammett and from Charles R. Dawson, then the presiding spirit over Columbia's undergraduate organic chemistry program. Except for an evening course I had taught at the Polytechnic Institute of Brooklyn while I was at U.S. Rubber, I had had no actual teaching experience, and I was somewhat worried about how I would make out. Happily, I found that I enjoyed teaching very much, and during my stint at Columbia, I regularly taught undergraduate organic chemistry; a graduate course in physical organic chemistry; and, occasionally, a special topics course on free radical reactions.

Columbia undergraduates were, for the most part, smart, enthusiastic, strongly motivated, competitive, and a pleasure to

Left to right: Charlie Dawson, Nick Turro, and Ronald Breslow. Together with Tom Katz, Gil Stork, and myself, they made up the senior organic staff at Columbia while I was there. The occasion is a house-warming party at the Storks' in 1968.

deal with. The quality of the graduate students was also high, and in retrospect, the 1950s and early 1960s were the golden age for teaching physical organic chemistry. Undergraduate organic chemistry in those days was largely descriptive, but students came out with an extensive factual knowledge of the subject. In a physical organic chemistry course it was then possible to show the principles and relations behind all these facts, and students found this approach very exciting. The present practice of presenting theory before and along with the facts, although it seems logical, doesn't stir up the same amount of enthusiasm, and higher level courses tend to be just more of the same in greater detail.

It was also a challenging time to be teaching the subject. New material was coming out all the time, so texts rapidly became out of date. I used to start my lectures by listing the recent original references I was going to talk about, and I encouraged the class to read them. Typically my course started out with a review of current ideas of structure, stereochemistry, acids and bases, linear free energy relations, etc., with a nod towards molecular orbital theory, which was really just appearing on the scene. We then considered what was known about the mechanisms of various classes of reaction with chief emphasis on nucleophilic displacements, solvolyses, carbonium ion rearrangements, carbonyl additions and, of course, radical reactions. Much of the material involved discussion of the effects of structure on reactivity and reaction rates, and I made an effort to present the actual numerical data. The stereochemistry of solvolyses and the concept of "nonclassical" carbonium ions were just unfolding in all their convoluted ramifications, and I struggled for years to make the picture comprehensible to students. Years later I tried to summarize my final views in a review titled "An Innocent Bystander Looks at the 2-Norbornyl Cation".[103]

As I see it, the peculiar geometry of the norbornyl system permits extensive delocalization of any charge on C-2 by interaction with the electron cloud around C-6, and this accounts for the experimental facts that the 2-norbornylcation is significantly stabilized, easily formed by solvolysis of 2-exo derivatives, and leads to rearranged products. This observation that carbocations can be stabilized in this way through non-bonded interactions is the important conclusion in regard to "nonclassical" ions. The question whether the 2-norbornyl cation is symmetric or consists of two isomers separated by a miniscule energy barrier in a way involves a special case and is much less important, although attempts to answer it have led to the development of some very elegant techniques, particularly the interpretation of the NMR spectra of carbocations, which have had fruitful application to other systems. Unfortunately the prolonged wrangling about these points was a real problem for editors and led to some disillusionment about the state of physical organic chemistry on the part of granting agencies.

Branching Out

Committee on Professional Training

The staff at Columbia took teaching very seriously, and I found that I became increasingly interested in the whole enterprise. In 1962 I was appointed a member of the Committee on Professional Training (CPT) of the American Chemical Society, I'm sure at the suggestion of Louis Hammett, who had previously been a long-time member. I remained connected with CPT until 1975, acting as chairman from 1964 to 1972. I found this a most re-

At an American Chemical Society meeting with John Bailar (left), 1972. Bailar was president of the American Chemical Society in 1959.

warding and satisfying experience. The committee worked hard, but it had splendid dinners at its semiannual meetings. Its membership included a number of distinguished chemists who have remained my close friends. During all the time I was with the CPT, much of its smooth functioning and effectiveness was due to its secretary, John Howard, whose contribution to American chemical education has never received the recognition it deserved.

The history and functions of the CPT have been described elsewhere.[104] Basically, it was set up to formulate (and periodically revise and update) the criteria for an adequate professional major in chemistry and to publish a list of those departments whose curricula met these criteria. In the post-Sputnik period when I was on the committee many new depart-

Right to left: Ed Wiig, our daughter Barbara, Mrs. Wiig, and Jane in Fairbanks, Alaska, June 1967. Ed for many years was an important member of the Committee on Professional Training of the ACS. When he retired at Rochester he went for a while to the University of Alaska and invited me to give a series of lectures there. It was quite a trip.

ments were applying for listing, and we were kept very busy evaluating them, a matter which I soon learned took a great deal of skill and tact. At present, the list stands at some 580 departments. The standards have provided chemistry departments with a powerful bargaining tool in dealing with their administrations and, I think, in large part, account for why chemistry is so often the strongest science department in an institution. Beyond this, the CPT publishes the *Directory of Graduate Research* and issues occasional reports on other topics concerned with chemical education.

The 1960s saw a great increase in Ph.D. production and the proliferation of new Ph.D. programs in chemistry. Two CPT reports during this period strongly urged that, whereas the aspirations of individual departments were laudable enough, the overall situation was clearly getting out of hand. New programs and facilities were outrunning the supply of potential students, not to mention the number of positions that would be available for them on graduation. These reports may have had some dampening effect, but the problem is clearly still with us, as shown by the increasing dependence of departments on students from abroad.

Research Groups

After some orientation lectures by the staff on the kinds of research in progress, our graduate students were encouraged to select a research sponsor. Because students were in short supply, those who did satisfactorily in our preliminary course work were rarely turned down unless a faculty member felt he was getting a larger research group than he could handle. In my own case, I would present one or more problems in which I was interested. After we had agreed on a topic and I had supplied some preliminary background and guidance, I largely turned the student loose without continued detailed supervision, although I would frequently ask how things were going and was always available for discussion. The students helped each other, and this scheme may not have turned out results as rapidly as possible, but it was generally successful. Almost all students produced publishable theses, and most have gone on to very successful careers.

During the time I was there, Columbia developed a very strong group in organic chemistry, which has made it one of the top departments in this area in the United States. Gilbert Stork was appointed at the same time as I and joined us in the winter of 1953, along with the first commercial IR instrument at Columbia, which he had insisted on for his research. He was shortly followed by Ronald Breslow and then Tom Katz and Nick Turro. Koji Nakanishi joined the department the year I left.

Relations between our groups were friendly and cordial, with students attending each other's group seminars, but I don't think that my colleagues' research ever had much impact on my own work. On the other hand, Ronald Breslow became interested in intramolecular radical reactions and has applied them very effectively in his "radical relay" technique for highly regiospecific reactions on steroids.[105] More recently, Gilbert Stork has made effective use of radical cyclizations of the sort that my group had demonstrated in the 1960s in his synthetic program.[106] This work has attracted well-deserved attention, although I've joked with him about the long time that had elapsed before this particular piece of technology transfer took place between adjacent floors of the Chandler Laboratory.

The only actual collaborative research I did at Columbia was with George Fraenkel. When I arrived, he was building one of the first high-sensitivity ESR (electron spin resonance) instruments in existence. With this equipment, we were able to observe the free radicals present in polymerizing methyl methacrylate, the first direct demonstration that free radicals were really involved.[107] One day at lunch, I remarked to him that the chain ends in molten sulfur were probably free radicals, and we spent the afternoon heating sulfur in a melting-point tube and sticking it in his instrument. We got a signal, which later turned out to be due to impurities and slap-dash technique. He interested a student in the problem, who found that, at higher temperatures, good spectra could be obtained, and the kinetics of the reactions occurring in the melt were determined.[108]

My time at Columbia also coincided with the great expansion of federal support of academic research, and I rode happily along with the tide. When I arrived, Louis Hammett arranged some funds from a DuPont grant-in-aid to the department to get me started and to purchase my initial high-pressure equipment. I soon obtained further support from the Office of Ordnance Research and the Office of Naval Research and, rather later, from the Petroleum Research Fund of the American Chemical Society.

Once the National Science Foundation (NSF) came into being in the late 1950s, it became the major source for my research support, supplemented by a variety of fellowships brought in by graduate students and postdoctoral fellows. Because my research group never exceeded 10 members, my needs have been modest. I have found the NSF to be both understanding and generous, sometimes giving me more and for longer periods than I had originally requested.

Other Involvements

For a time in the 1950s, through the urging of Morris Kharasch, I became involved with some research for the Chemical Warfare Service. This was a peculiar arrangement, because the work was initially classified, with all the attendant complications. For staff, they supplied me with two very competent Ph.D. draftees, Earl Huyser and Frank Stacey, who found Columbia a great improvement over raking leaves at an army post. They served in uniform, and we referred to them, jokingly, as my armed guard. Although the work, a study of the cooxidation of hydrocarbons and elemental phosphorus as a possible route to organophosphorus compounds, was eventually published,[109] I was never very happy with the arrangement, which I felt was inappropriate for a university, and I was glad to see it terminated.

When I left Lever Brothers Company, I was retained by them as a consultant. Because both polymer chemistry and radical chemistry are important to industry, I soon found myself involved in a succession of other consulting arrangements, largely with petroleum companies, as well as giving lectures or short courses at industrial laboratories. The longest lasting consultantship was with the Celanese Corporation, subsequently Hoechst-Celanese, which began in the 1950s and continued until my retirement. The work with Celanese is also the only instance in which I made a significant contribution to an important product, the plastic Celcon, a formaldehyde polymer stabilized by copolymerization with a small amount of ethylene oxide.[110]

The concept arose as a suggestion I made at a research conference. The next day I decided that it was sufficiently important that I write Celanese a letter to make sure that it was not forgotten and that some work would be done. Happily, the concept worked, the polymer was sufficiently stabilized to be a commercially desirable product, and the letter later proved useful in legal arguments between Celanese and DuPont, who was marketing a competitive product.

The chemistry involved is rather neat. A high-molecular-weight polyacetal of formaldehyde can be prepared in several ways and has hemiacetal end groups unless they are protected in some way. Because monomer and polymer are in potential equilibrium, the chains can be "unzipped" in the presence of mild acid or base or by simple heating

$$-O-CH_2-O-CH_2-O-CH_2-OH \rightarrow -O-CH_2-O-CH_2-OH + C_2H{=}O \text{ etc.} \quad (30)$$

In the copolymer $-O-CH_2CH_2-O-$ units are distributed at random along the chain, and the unzipping stops at this point because the end group is now a stable acetal.

$$-O-CH_2-O-CH_2CH_2-O-CH_2-OH \rightarrow -O-CH_2-O-CH_2CH_2-OH + C_2H{=}O \quad (31)$$

If a chain is somehow broken in the middle, unzipping is similarly limited. The $-O-CH_2CH_2-O-$ units decrease chain regularity and yield a slightly softer and less rigid polymer with a lower softening point, which, in turn, make the product easier to mold.

I found consulting time-consuming (for a long period, I was spending almost a day a week at it) but worthwhile, not only in supplementing what was then a rather modest academic income, but also in keeping me informed of industrial problems. For years we have been hearing about the importance of technology transfer. Many academic chemists are remarkably uninformed about what goes on in industry, and I've long felt that a greater use of academic chemists as consultants could be a relatively simple, inexpensive, and mutually beneficial way of facilitating mutual understanding.

Consulting relationships between academia and industry can go both ways. In about 1966 when the hazards of cigarette smoking were receiving much attention, a department chairman at the Columbia Medical School was approached by an entrepreneur who claimed to have developed a cigarette filter that was miraculously effective in removing tar and nicotine from cigarette smoke. His proposition was that you can't stop people from smoking, but you can make the smoke relatively harmless. If Columbia would vouch for the efficiency of the filter, the entrepreneur promised to split the royalties with the university. The Columbia administration bought the idea and announced the deal at a press conference that I had gotten wind of and attended.

I was concerned about who was advising the administration, so I called the university treasurer who was handling matters. I told him that as a consultant to Celanese, a major producer of cigarette filters, I had access to information and would be happy to help. He thanked me, and I went on to learn from Celanese the obvious fact that the efficiency of a filter depends both on its composition and on how tightly it was packed. Too tight packing leads to such a large pressure drop through the cigarette that it cannot be smoked.

A few days later I got a call from the treasurer saying that the project was in trouble. On looking over the data that had been supplied to Columbia, there were plenty of figures on tar and nicotine removal but no mention of pressure drop. I visited the testing laboratory where the data had been gathered and found that the cigarettes had been supplied by the entrepreneur, and no measurement of pressure drop had been requested or made. When this was done, the cigarettes with the miraculous filters were indeed unsmokable, and the miraculous material, when packed to give the same pressure drop, had no advantage over standard filters on the market. I was able to convince the Columbia administration that they had purchased a worthless pig in a poke, but I'm afraid it cost them considerable time and money to extract themselves from the agreement.

There were academic entrepreneurs in those days too, and some of my friends with pharmaceutical connections did very well at it. My only adventure in that direction at least makes a good story. When I went to Columbia we moved back to Montclair, New Jersey, where I had a neighbor named Phil Ruppert. During our exam period in 1955 Jane and I joined the Rupperts in Antigua, then part of the British West Indies. I found Antigua was a beautiful place to correct examinations, although snorkeling around a coral reef was really more fun.

Phil worked for the National Lead Company but was quite an entrepreneur on the side. He persuaded me and another friend of his that Antigua would be an ideal place to grow castor beans. An enterprise named Antigua Castor Enterprises came into being, which agreed to supply the island's agricultural administrator with castor beans to plant, and then we would buy the crop. This wasn't as harebrained as it sounds—the Baker Castor Oil Company was a subsidiary of National Lead, so we had both technical advice and an assured market. Unfortunately, in the Antigua climate the plants produced more leaves than beans, and most of the beans we did get mysteriously went to England.

At this point the tourist boom in Antigua was just starting, and we decided that tourists might be a better crop than beans. Antigua Castor Enterprises sprouted a subsidiary chartered to operate a hotel, and shortly one appeared on a rise of ground overlooking a lovely beach and bay. It had a central lodge, swimming pool, dance floor, and cottages. To find your way to the cottages there were knee-high lights under which, at night, chicken-sized toads squatted, catching moths. What followed was a series of problems: the location was so romantic that the manager's wife eloped with the chef the week after we opened. Next, the rum was so good we had to fire the second manager. Finally we got a couple to manage the place, but, under the spell of the tropics, both departed with other partners. Clearly the project was going to take more attention than we could give it, so we sold it to a new establishment next door (as a final inducement to buy, Phil arranged to have our steel band play all night at their corner of the property). Unlike most neophytes in the business we didn't actually lose any money, but about broke even.

Travel

During my career I've seen travel become very much one of the perks of academic life and have come to feel that we now have rather more meetings than there are new and significant results to talk about. However, although I haven't tried to compete with some of my roving colleagues, much of the traveling I have done has been highly rewarding. I've mentioned the Faraday Society meeting in 1947. In 1958 I was invited to spend a few summer weeks at the University of Washington in Seattle, so Jane and I loaded our four oldest children into an overburdened Ford station wagon and set out on a 10,000-mile tour of the West. Seattle was pleasant, and the chemistry faculty, particularly Ken Wiberg, who had not yet moved to Yale, was stimulating. The chief benefit, though, was the prolonged family outing and the chance to show the children national parks, the West Coast, a dude ranch, and other western attractions.

Our trip west in 1958 was not without its problems. Here we see our four children changing a tire in the Nevada desert.

At the Harvard Commencement 1947: Generals George Marshall (holding the hat) and Omar Bradley (in uniform). This was my 10th reunion, and I managed to take this picture. It was a historic occasion. Marshall, Bradley, and Robert Oppenheimer all received honorary degrees, and at the alumni meeting in the afternoon Marshall announced the Marshall plan. I'm afraid that most of us in the audience didn't realize its full significance at the time.

Leopold Ruzicka (left) and Vladimir Prelog at the E.T.H. in Zurich in 1960. I first met these two distinguished organic chemists in my visit to Europe that year.

In 1960 I enjoyed my first sabbatical leave. After some consulting trips, Jane and I with Frank and Ellie Mayo left on a tour of European laboratories, ending up at the polymer symposium in Moscow mentioned earlier. On our arrival in England Frank took delivery on a magnificent new Jaguar sedan, so we toured Great Britain, Holland, Germany, Switzerland, Austria, and a few other countries in considerable style, visiting more than a dozen universities and industrial laboratories and still finding time for some sightseeing. The car proved a good investment for Frank—he drove it for the rest of his life.

We got back at the end of June, and, after catching our breath, Jane and I set out with four of the children for Durango, Colorado. Here Paul Bartlett had agreed to run a National Science Foundation (NSF) sponsored summer course on physical organic chemistry for college chemistry teachers at Fort Lewis A&M College. He had recruited me and persuaded Jack Roberts and George Hammond to divide the term between them. We all had a wonderful time. There were two sessions a day with afternoons free and weekends available for trips. We all attended each others' lectures and felt free to make comments and ask

At the National Science Foundation Summer Institute in Durango, Colorado, 1960. Left to right: Jack Roberts, Paul Bartlett, and Edith Roberts.

Ed King (right) and I at the top of Hallett Peak in Rocky Mountain National Park in 1965. This was the summer I spent in Boulder at the University of Colorado. Ed was a long-time colleague of mine on the Committee of Professional Training of the American Chemical Society.

Left to right: me, Sadie Walton, and Jane at a joint meeting of the American and Mexican Chemical Societies in Mexico City in December, 1975. Sadie's husband, Harold Walton, was department chairman when we spent the summer at the University of Colorado in 1965, and we all became very good friends.

Fred Hawthorne and plane, Riverside, California, 1967. Among other things, Fred did important work in boron chemistry. He was a colorful character who liked to give pyrrophoric displays with his boron compounds and was an enthusiastic amateur pilot. One of the high points of my winter in California in 1967 was going up for a flight with him.

questions. The participants, about 30 in number, seemed to enjoy it too. They felt they had learned a lot, and I've continued to run into them at meetings and elsewhere. Most NSF courses of this sort have been run for high-school teachers, and this was one of the few at the college level. In view of its success, I'm sorry there haven't been more.

In 1965 we had another fine summer when I taught a summer course at the University of Colorado in Boulder, although by this time our crew was reduced to my two youngest children, the others being married or otherwise occupied. Again I enjoyed my associates and the countryside. We ended up at a ranch outside of Estes Park, and the last day I arranged for a guide to climb Long's Peak with me. We started well before dawn, and when it got light I was disconcerted to find that he was a high-school boy from Seattle. Fortunately, he knew much more about mountain climbing than I did and got me to the summit and back without incident and with splendid memories of the view from the top.

The University of Utah, 1969 to the Present

In 1969 I moved from Columbia University to the University of Utah. My reasons were complex, and to my Columbia colleagues, I'm sure, I appeared quixotic. I had always liked the West, and its attraction had been increased by some months that I had spent in California while on sabbatical leave in 1967. My attachment to Columbia had been shaken by the student demonstrations there in the late 1960s. Although their causes were understandable, I felt that no one—students, faculty, or administration—had distinguished themselves. My research problems were nearing completion; I was tired of commuting from Montclair, New Jersey, to New York; and because four of my five children were grown and had left home, our house was too large. Really, I'd gotten restless and felt the need for a change.

I had tentative conversations with a number of universities, and my choice of Utah, aside from its geographically attractive location, was mostly the result of the effective persuasion of James Fletcher, then president of the university, and David Grant, the department chairman. In particular, Dave had put together a remarkably well-thought-out proposal to NSF for a "Center of Excellence" grant, which was being funded, and the prospects for rapid departmental growth and development looked good.

I'm happy to say that these expectations were largely realized. At the time I came, Dave had also recruited R. W. Parry from Michigan and several younger staff members. Chemistry had just moved into a large, well-equipped new building to which a substantial addition was subsequently added. The department has had strong continuing administrative support. Faculty recruitment has continued to be remarkably successful, and I'm proud of the support we have been able to give our

Bob Parry, me, and Henry Eyring (left to right) at the University of Utah in 1970. Bob and I had recently arrived. Henry, as long-time dean of the college of science, was largely responsible for putting the university on the map scientifically.

younger faculty members. I think Utah now ranks firmly in the second group of strong departments, immediately behind the small set of prestigious establishments like Harvard, Berkeley, and Columbia.

Nevertheless I found that there were real differences. As a state university, Utah has an easygoing admission policy. There was less intellectual curiosity among the students and undergraduate courses were less exciting to teach. If I could make a generalization, it is that although many of the students were smart enough, they didn't realize the effort required to do a really good piece of work. The ratio of graduate students to staff was smaller, and I relied chiefly on post-doctorals for my research in part to avoid competing with the younger staff for the limited supply. During my time there I taught various sections of undergraduate organic chemistry and graduate courses in physical organic chemistry and in kinetics. In kinetics I experimented with problems based on individual computer-generated data (enter your social security number and it is converted into rate constants, which then generate data, complete with experimental error; your problem is to figure out what the constants are).

I also periodically gave a special-topics course on free radical reactions (a splendid way of forcing myself to keep up on current literature) and, a few times, an experimental course on science, technology, and society. This amounted to a survey of the parallel historical development of chemical science and technology, ending up with a discussion of current problems, largely environmental. As part of the course (and to give them an idea of how little they knew of the real world), I would ask students to imagine that they'd been sent to a third-world country with instructions to set up a chemical industry relying on local resources and to consider what would they do. It really baffled them. The course included a discussion of the problem of halocarbons and the ozone layer. Here I could draw on what I had learned as a member of the first National Academy of Sciences committee that looked into the matter in 1976–1979.

In spite of its location, I found Utah remarkably unisolated. Usually we had a steady stream of visitors with several outside speakers at seminars a week and had little trouble attracting people for longer periods to give courses or lecture series. In the

Triple Arch, Arches National Park, Utah, 1980. The figure in the foreground is Jean-Marie Lehn. He gave an elegant series of lectures at the University, and I took him on a tour of some of Utah's natural wonders.

winter, skiing was certainly an attraction, and in the summer, Utah's mountains and canyons were available.

I certainly made ample use of both. The ski resorts were a half hour or so from my house, and in the summer the Wasatch Canyons were full of hiking trails leading to little mountain lakes. For longer trips there were the national parks and back country like the Henry Mountains. I became quite adept at coaxing a four-wheel-drive car over vestigial roads. All in all I found

My son Cheves (left) and I on the summit of Mt. Ellen (11615 ft), the highest peak in the Henry Mountains of Utah, 1972. On the slope in the background could be seen (through field glasses) the only free-ranging herd of buffalo in the United States.

Utah a very pleasant place to work and enjoyed my many congenial colleagues. Still, because of many friends and children in the East we continued to maintain a place in New Hampshire and to spend at least part of our summers there.

Chemically Induced Dynamic Nuclear Polarization

Although at Utah we continued some of our peroxide and hypochlorite investigations, the rest of my research took rather an abrupt turn. One new topic was the application of chemically induced dynamic nuclear polarization (CIDNP) to the study of radical reactions. I had noted reports of this new phenomenon while at Columbia and was, admittedly, quite mystified about it myself, but I was unable to get anyone to work on it.

At Utah I was joined by A. R. Lepley, a visiting professor on leave from Franklin and Marshall College. Lepley had done some of the early work and was familiar with the developing theory. We also had available the facilities and advice of Dave Grant, a recognized expert in NMR spectrometry. With Lepley and, later, E. M. Schulman, we published several papers, including an attempt to use the intensity of CIDNP polarizations to measure the rates of radical reactions[111] and one of the first examples of ^{13}C CIDNP.[112] Although these may not have been major advances, I at least ended up with some understanding of the phenomenon and an appreciation of how it might be used.

Hydroxyl Radical Chemistry

A more long-lasting project, which went through several phases, started with an investigation of hydroxyl radical chemistry, initially to compare the properties of hydroxyl radicals with those

Ed and Dianne Schulman in Salt Lake City, 1973. Ed was a post-doc with me and became my expert on chemically induced dynamic nuclear polarization (CIDNP). The occasion was a farewell party for them before he left for the University of South Carolina.

of alkoxy radicals, which we had studied at such length. In looking for an approach, I recalled some seminal work by Merz and Waters,[113] of which I had learned at the 1947 Faraday Society discussion over 20 years before. Waters's approach employed Fenton's reagent, ferrous ion plus hydrogen peroxide, and involved simple addition of a known amount of H_2O_2 to a system containing Fe^{2+} and an organic substrate and determination of the Fe^{2+} oxidized.

The postulated reaction steps, slightly modernized, are given in Scheme 6.

$$
\begin{aligned}
Fe^{2+} + H_2O_2 &\longrightarrow Fe^{3+} + OH^- + HO\cdot \\
Fe^{2+} + HO\cdot &\longrightarrow Fe^{3+} + OH^- \\
HO\cdot + RH &\longrightarrow H_2O + R\cdot \\
Fe^{3+} + R\cdot &\longrightarrow Fe^{2+} + \text{product} \\
2R\cdot &\longrightarrow R{-}R \\
Fe^{2+} + H^+ + R\cdot &\longrightarrow Fe^{3+} + RH
\end{aligned}
$$

Scheme 6.

Intermediate radicals fell into three classes: those easily oxidized by Fe^{3+}, those reduced by Fe^{2+}, and those inert to either. By simply following reaction stoichiometry (the ratio of Fe^{2+} to the H_2O_2 consumed) and by properly plotting data, the radical type could be identified and the relative reactivities of RH and Fe^{2+} toward $HO\cdot$ could be determined. I've always liked simple experiments, and I've enjoyed pointing out that most of our work on hydroxyl radicals was carried out with no more equipment than a burette and a few flasks. It is important, though, to have a rational model and know what you are doing.

Our investigation was begun by a very able visiting Japanese colleague, Shin'ichi Kato,[114] and was continued by Gamil El Taliawi, a student who had started with me at Columbia and is now a professor in the University of Cairo, and a series of postdoctoral fellows.[115] The results gave a good picture of substrate reactivities, and we were able to show, for example, that, with isopropyl alcohol, some 15% of the reaction occurred via HO· attack on the methyl C–H bonds.

This work attracted considerable attention, because since Waters's original work, iron-catalyzed peroxide reactions had become of great interest to biochemists as models of enzyme processes, and an alternative formulation of the Fenton reagent reaction had developed, involving a higher valence iron species, the ferryl radical (formed via equation 32), which, in turn, attacked the substrate.

$$H_2O_2 + Fe^{2+} \longrightarrow FeOH^{3+} + OH^- \qquad (32)$$

Although the two schemes are kinetically equivalent, we showed that Fenton reagent selectivities for different substrates were in substantial agreement with HO· selectivities measured by radiation chemists in the absence of iron and that several other arguments in favor of the ferryl radical scheme were, in fact, fallacious. With Andre Goosen,[116] we were also able to resolve a similar controversy about the mechanism of the (slower) Fe^{3+}-catalyzed decomposition of H_2O_2 (equation 33), for which both a hydroxyl radical chain mechanism and a ferryl radical scheme had been proposed.

$$2H_2O_2 \xrightarrow{Fe3+} 2\,H_2O + O_2 \qquad (33)$$

The reaction was known to be retarded by organic substrates, and we were able to show that the magnitude of the retardation, for several substrates, corresponded well with that predicted by the HO· scheme and the known rates of HO·–substrate reactions.

Although we were able to extend our model to Fe–EDTA–substrate systems,[117] we clearly recognized and stated that, with other ligands attached to iron, higher valence species could well be the primary oxidants involved. Indeed, in many enzyme systems and models employing iron porphyrins with varying ligands, that now seems clearly the case. Products and reaction paths are quite different, but the Fenton reagent data provide an important basis for comparison.

Radical Cations

As our Fenton reagent work developed, the behavior of aromatic substrates provided a serious puzzle. According to the literature, the reaction of aromatic substrates sometimes resulted in ring hydroxylation and, at other times, in side-chain oxidation. For this problem, the key was provided by the proposal of R. O. C. Norman,[118] based on ESR spectra, that hydroxycyclohexadienyl radicals, formed by HO• addition to aromatic substrates, could undergo an acid-catalyzed dehydration to radical cations, which, by side-chain cleavage, could lead to the side-chain oxidations.

With this clue to guide experiment and the recognition that $SO_4{\cdot}^-$ radicals, derived from peroxydisulfate, directly oxidized aromatic substrates to radical cations, we were able to show that a variety of data were consistent with a scheme, shown in Scheme 7 for toluene. In this scheme, OX represents Fe^{3+}, Cu^{2+}, or other oxidants.

Interestingly, isomer distributions of phenolic products were found to vary significantly, depending on whether the rate-controlling step in hydroxycyclohexadienyl radical formation was hydroxyl radical addition or hydration of the radical cation, and the relative rates of oxidation of the possible isomers.[119]

A very important analog of the reversible hydration of aromatic cations shown in Scheme 7 occurs in suitable aliphatic systems. The ESR spectra of systems involving HO• radical attack on ethylene glycol show evidence for the $\bullet CH_2CH{=}O$ radical, and it had been proposed that it arose from an acid catalyzed β-OH loss from the expected radical to give what might be called the radical–cation of acetaldehyde enol followed by proton loss to give the radical observed. Because the interpretation was based solely on ESR spectra, Richard Johnson examined the reaction of ethylene glycol with Fenton's reagent and showed that both stoichiometry and isolated products were consistent with Scheme 8.[120]

The dehydration competes with Fe^{3+} oxidation of the initial radical and is very rapid, with a rate constant in 0.5 M acid of about 1.3×10^8 sec^{-1}. 2,3-Butanediol behaves similarly and

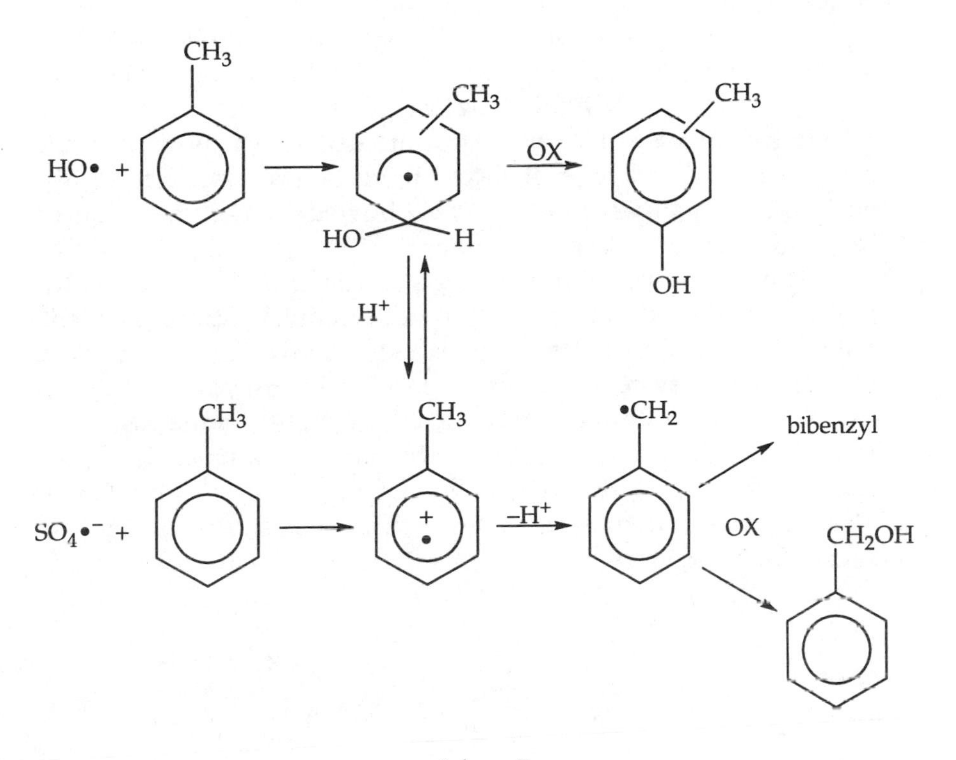

Scheme 7.

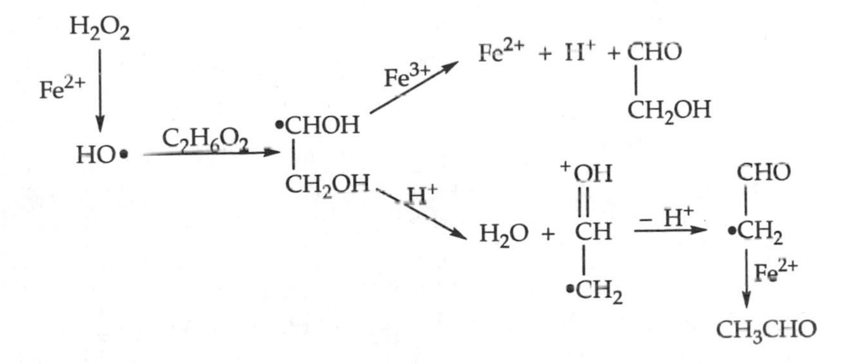

Scheme 8.

other good leaving groups such as Cl or phosphate in the β-position to the radical are lost even more readily.

This sort of reaction has subsequently received considerable study and provides a model for a number of important biological processes. Several enzyme-catalyzed rearrangements involving coenzyme B_{12} can be formulated this way as can the oxidative radical cleavage of DNA. Here radical attack at a suitable point on a sugar unit may be followed by β-elimination of phosphate leading to chain cleavage.

The final turn of this rather protean project has been toward the further chemistry of aromatic radical cations and the modes of cleavage of more complex side chains. For this work, a good oxidizing system has proved to be peroxydisulfate plus Cu^{2+} (to oxidize intermediate radicals) in slightly aqueous acetic acid or acetonitrile (to improve substrate solubility). Our most recent work, carried out chiefly by Chengxue Zhao and Gamil El Taliawi, has shown that at least three kinds of cleavage are possible (Scheme 9) depending on side-chain structure.

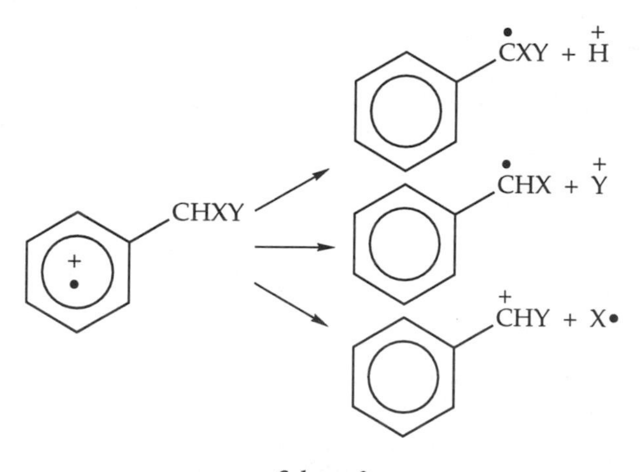

Scheme 9.

Reaction paths vary with conditions, and because intermediate products may be further oxidized, a remarkable variety of products is sometimes formed.[121] For example, in the rather simple case of cumene, we observed the reaction sequence shown in Scheme 10.

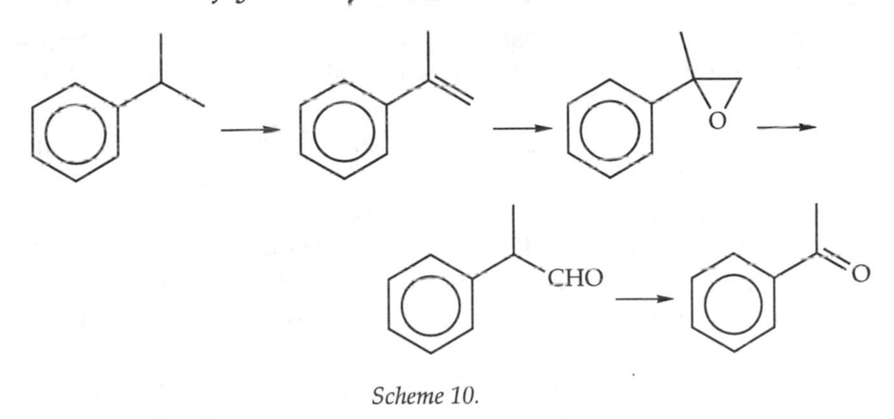

Scheme 10.

Radical cations are now recognized as intermediates in a great variety of oxidations, and interest in their reactions is growing. Although our work in the field is finished, I look forward to future developments in the hands of others.

My Adventures as an Editor

In the summer of 1974, I received a call from Bryce Crawford, then on the board of directors of the American Chemical Society, asking if I would like to be considered for appointment as the editor of the *Journal of the American Chemical Society* (JACS). I had first encountered the JACS as an undergraduate at Harvard. Arthur Lamb was then editor, and I regularly passed his rather mysterious editorial office every day on my way to class. Subsequently, I had published most of my own work in the JACS; had been a quite regular referee; and, in the 1960s, had served on its editorial advisory board. Nevertheless, Crawford's inquiry came as a complete surprise, and I was both flattered and alarmed. After several meetings in which I got some idea of what was involved, I accepted and assumed the editorship on January 1, 1975. Fortunately, I had been able to acquire a very competent editorial secretary, Charlotte Sauer, and two of my colleagues, Dave Grant and Bob Parry, agreed to serve as two of my associate editors. Bob, who had had extensive editorial experience and had been editor of *Inorganic Chemistry,* was a valuable source of advice. I recall that when I asked him what being an editor involved, he said, "It is like standing under a waterfall."

Bob's simile proved to be apt. During the 7 years, 1975–1981, that I held the position, my office received over 3000 manuscripts a year. Although most papers were referred to associate editors, I handled 400–500 manuscripts a year myself. I found the operation fascinating, but it certainly preempted a good deal of the time I would have devoted to research and other activities.

As the time to start editing approached, I tried to assemble my ideas on editorial policy and wrote up a short statement of my views,[122] a practice that I continued with similar reports in subsequent years. As I saw it, a publication like the JACS performs three functions. First, and rather uniquely, it is a permanent repository of the ideas and the data on which the edifice of science is erected. Second, it is a news sheet informing readers on what is currently being done and thus competes with meetings, private correspondence, and other means of exchanging information. Third, it provides authors with a forum for their ideas and a medium for advancing their position and

With Fred Greene (left) at a meeting of American Chemical Society editors, 1978. At that time I was editor of the Journal of the American Chemical Society *and he was editor of the* Journal of Organic Chemistry *and a professor at M.I.T.*

With Charlotte Sauer at a lab party at the University of Utah on my birthday, 1989. Charlotte was my extraordinarily competent editorial assistant during the first years that I was editor of the Journal of the American Chemical Society *and has continued in the same capacity with Peter Stang when he became associate editor.*

standing. For readers and the scientific community as a whole, the importance of these functions probably lies in the order indicated, but for authors, the order tends to be reversed.

An editor finds himself in a position requiring a good deal of judgement, tact, and sometimes firmness. Sometimes, he may get the feeling that he is the readers' only defense against ambitious authors. He must, however, retain the authors' good will and confidence or he may have nothing to publish. I found it an exhilarating balancing act and concluded that the best thing to do when I made a mistake was to admit it promptly. Overall, the successful operation of a journal is a cooperative effort on the editor, authors, and referees, and I must put in a good word for this last group. Competent refereeing is a real load on the profession in terms of time and effort and not only helps to eliminate errors and trivia but also often greatly improves manuscripts. Good referees perform an invaluable service even though the final decision lies with the editor.

In regard to refereeing, I once conducted a slightly tongue-in-cheek survey that I never published, although I showed it to my advisory board and contributed a copy to an NSF discussion of peer review. In it I took the first 200 or so Communications that had been submitted one year for which I had two clear-cut referees' reports and counted the number with two favorable reports, those with two reports recommending rejection, and those with split decisions. I then analyzed the data on three assumptions:

1. Referees' reports were purely random. Conclusion: unlikely; the number of split decisions was too small by about two sigma.

2. There are three kinds of referees: those who accept everything, those who reject everything, and those who can tell good from bad. Conclusion: more than half of the referees are reliable.

3. Referees are all right but there are three kinds of manuscripts: good, bad, and so marginal that what a referee says is a toss-up. Conclusion: about half are borderline.

As we know, the real world is more complicated, and all these factors can enter in. Besides, an editor who knows his referees can influence how the reports come out. He may choose a referee who he is sure will recommend rejection because he wants his opinion on a specific point. If he does this, though, he must be sure to take it into account in making his decision.

A second editorial problem is simply the matter of numbers. The page budget of the JACS permitted (and still permits) it to publish about 60% of the manuscripts submitted. Some papers are clearly outstanding, some are trivia or wrong, but the majority are in a middle group—of interest to some readers but not more significant than many others. Out of these, an editor must make a selection, and as I've told authors, editing a scientific journal is much more like editing a poetry magazine than they realize. The goal, as I see it, is largely to give readers an interesting mix, and this requires a great many arbitrary decisions: a prolific author should send some of his less exciting work elsewhere; here is something not earth shaking but still novel; this author deserves an audience; or this field has been worked over to exhaustion. Finally, perhaps disappointment should be spread as evenly as possible.

A third problem is the matter of detail. The largest group of readers simply scans a paper for its conclusions. A somewhat smaller group is interested in the logic and general types of data on which these conclusions are based. Finally, a few readers want to check the theory and data in detail or to use the experimental techniques in their own work. Clearly, to supply all readers with everything about a piece of work is expensive and wasteful, but these details need to be available somehow.

When I became editor, the Books and Journals Division of the ACS was concerned with this problem and proposed that the JACS might be published in two forms: as long abstracts, which would be subscribed to by most readers, and in an archival form with full papers, chiefly for libraries. I was glad to cooperate with a trial. With the help of a number of authors, several experimental issues were put together in this form and widely circulated. Reactions were mixed but not sufficiently enthusiastic to justify the change. At present this problem remains unsolved. Perhaps the best solution, with the least change, would be an expansion of the use of supplemental material (typescript, at present available as microfiche), which the JACS presently offers, particularly if it could also be made available for computer retrieval. I'll admit I am somewhat alarmed by the current discussion of publishing journals by simply inserting manuscripts into the computer network. Seeing the gossip that flies around on e-mail, without adequate refereeing we would be deluged with half-digested trivia, and with it I'm not sure how much time would be saved. If things are too easy, no one will do them carefully.

My 7 years as an editor sometimes seemed to go on for a long time, but it was an experience I wouldn't have missed. Subsequently, I was rediscovered as a referee, so I could still keep a hand in the game. When I took the job, the JACS was probably the premier U.S. journal in most areas of chemistry. Science has become more international, and the JACS now gets some 30% of its manuscripts from abroad. I like to think its premier status is now worldwide, and I'm proud to have been associated with it.

Other Adventures

My career at Utah also seemed to increase my traveling. On top of the usual round of meetings and conferences, including a number abroad, in 1974 Jane and I spent 2 weeks in Egypt at the invitation of Gamil El Taliawi on a trip that combined lecturing at the University of Cairo with a tour of antiquities extending from Alexandria to Aswan.

In 1978, thanks to a fellowship from the Japanese Society for the Promotion of Science, we spent a splendid 6 weeks in Japan in which we covered the islands from Sapporo to Mt. Aso, combining sightseeing with visiting most of the major universities. Our trip was arranged by Hideki Sakurai at Sendai and Shigaru Oae at Tsukuba, both of whom I had met at a Japanese–American conference on peroxide chemistry in Boulder, Colorado, in 1976, and Japanese courtesy and hospitality outdid themselves. I think we were particularly charmed by Kyoto.

Egyptian nightlife on our visit there in 1974. Jane and I with Gamil El Taliawi (left) at a nightclub in a tent in the desert near Cairo.

Japan, 1978. Jane and I with a Samurai warrior in full armor (he's stuffed).

Tsukuba, Japan, 1978. Me with W. Ando (on my right) and the proprietors of a Japanese stone lantern factory. Ando is helping me buy the stone lantern (rather smaller than those in the background) that now graces our garden.

With Hideki Sakurai (right) at our house in Salt Lake City, 1979. Hideki had organized much of our trip to Japan in 1978. He was now visiting the University of Utah, and we were struggling to equal Japanese hospitality.

Lanzhou University, People's Republic of China, 1982. Professor You Cheng Liu is on my right, and the poster announces the series of lectures I am going to give (I had it on my wall at Utah until it faded away).

Chengxue Zhao (left) and myself on the Great Wall of China on a cold October day in 1982. Chengxue had been a visiting scholar at Utah and accompanied us on our tour of China that year.

At a luncheon attended by physical organic chemists in Washington, DC, in August, 1983. Left to right: Jeff Seeman, Ernest Eliel, me, Tony Trozzolo, Paul Bartlett, Frank Westheimer, and Stan Tarbell.

In 1982 I was invited by the Chinese Academy of Sciences to visit the People's Republic of China, and we spent 3 weeks there, starting in Shanghai and getting as far west as Lanzhou, where Y. C. Liu had been doing free radical chemistry under considerable difficulty for many years. Chengxue Zhao had just returned to Shanghai from Utah, and we were fortunately able to arrange to have him accompany us on our travels. China was fascinating, but I found lecturing with almost simultaneous translation (I would say a few sentences and then drink tea while they were being translated) a bit wearing and admired my audiences' stamina. At the end of the trip we were able to go on to Australia via Hong Kong. Jim Morrison at Latrobe University, a mass spectrometrist who had been a frequent visitor at Utah (where he held an adjunct appointment), arranged a whirlwind tour for us to Melbourne, Adelaide, Canberra, Sydney, and Brisbane with a chance to catch our breath at Heron Island. We'd heard about Australian beer, but came to appreciate their wine as well, which seemed to flow like water and certainly whetted our thirst to return.

The chance to do this came in 1987. I'd raised the possibility with A. L. J. Beckwith at the Australian National University in Canberra at the Free Radical Conference in St. Andrews, and he arranged for a 2-month visit there. True, I gave some lectures and discussed a good deal of chemistry (Beckwith had taken over and greatly expanded our early work on radical cyclizations using tin hydrides), but we chiefly enjoyed ourselves with trips to the beaches and the mountains and to Melbourne to see the Morrisons. Many things about Australia seem British, but everything is a little different, which makes it fascinating. Birds are a different color, there are kangaroos on the golf courses, and at picnics the problem is to keep the emus from eating the lamb chops. I was so beguiled that I wrote a poem on Australian fauna duly published in the *Research School of Chemistry News* of the Australian National University. Because it is my only public attempt of this form of expression, I'll quote a stanza:

A miniature marsupial bear,
The WOMBAT slumbers in his lair,
Or, waking, prowls the wild outback
Shaped like a furry flour sack.

We spent the winter of 1987 with Athel Beckwith at the Australian National University in Canberra. The next year I received this cheery greeting from him (center row) and his research group. Andreas Zavitsas, who took his degree with me about 1963, was visiting from Long Island University, New York, and is on the extreme left.

Our visit ended with a celebration honoring Arthur Birch, at which Gilbert Stork was a principal speaker and I participated with a largely historical talk. Gilbert had to return to Columbia, but his wife, Winifred, then came with us on the train trip from Sydney to Perth (one of the few spectacular long train rides left in the world), returning by plane via Ayres Rock, another Australian wonder.

"Cold Fusion"

I'll end my more or less chronological account with some comment on the bizarre and frustrating project on which I spent my last 2 years at Utah. The year 1986 marked my 70th birthday, for

Tito Scaiano (left) and Wes Bentrude at the free radical symposium in Zurich, 1988. Wes was a colleague of mine at Utah and an expert on the radical chemistry of organophosphorus compounds. Tito (now at Ottawa) has done phenomenal work on measuring fast reaction rates by fast laser photolysis.

which my colleagues kindly organized a symposium titled "Free Radicals in Perspective". They coaxed a number of my former students and old friends to attend, including Frank Mayo and Keith Ingold, who acted as the principal speaker at the dinner. It is hard not to enjoy and be moved by such an occasion, and I certainly was, particularly because all my children were able to be there. By this time, my laboratory research was wound up, and I shortly went on to a half-time appointment with the announced intention of retiring completely at the end of the 1988–1989 academic year.

Shortly before Christmas, 1988, Stanley Pons, then our department chairman, mentioned to me that he thought he had detected a nuclear reaction during the electrolysis of D_2O in a cell with a palladium cathode. I was intrigued and astonished and, during the winter, occasionally heard more, including that tritium formation had been detected and that there was a developing rivalry with Stephen Jones and his group at Brigham Young University who were said to have made similar observations. I became increasingly interested, particularly when in mid-March Pons showed me a preprint of the paper that he and Martin Fleischmann had submitted to the *Journal of Electroanalytical Chemistry*[123] and told me that there was to be a public announcement and news conference on March 23.

The first paragraph of the accompanying press release set the tone of this now notorious conference in which Chase Peterson, the president of the university, and James Brophy, the vice president for research, took as prominent parts as Fleischmann and Pons: "Two scientists have successfully created a sustained nuclear fusion reaction at room temperature in a chemistry laboratory at the University of Utah. The breakthrough means that the world may someday rely on fusion for a clean, virtually inexhaustible source of energy."

The centerpiece of the announcement was a simple electrolytic cell with a central palladium cathode and a platinum

In Yosemite National Park, California, in 1979. Left to right: Ellie Mayo, Frank Mayo, and me.

At the International Symposium on Free Radicals in Freiburg, Germany, 1981. Left to right: Chris Foote (University of California, Los Angeles), me, and Shigaru Oae (then at Tsukuba University).

At an outing during the 3rd International Symposium on Organic Free Radicals in St. Andrews, Scotland, 1984. Left to right: Jane Walling, Keith Ingold, me, Athel Beckwith, Kaye Beckwith, and Cairine Ingold. Keith, at the National Research Council of Canada, was a leader in the measurement of the rates of fast radical reactions and a long-time friend.

Speakers at my 70th birthday symposium, Park City, Utah, 1986. Left to right: Keith Ingold, Wes Bentrude, Peter Wagner, Athel Beckwith, Bob Neuman, me, Dennis Tanner, Al Padwa, Jay Kochi, Gilbert Stork, and Ned Porter. Al Padwa had kindly provided the T-shirts with free radical insignia.

Jane and I with most of our children at the symposium in Park City, Utah, in April 1986 celebrating my 70th birthday. Left to right: me, Hazel, Cheves, Barbara, and Rosalind. Jane is seated. Our fourth daughter, Janie, was detained by bad weather but made it for the post-symposium skiing.

wire helical anode with D_2O–LiOD as electrolyte, also containing a temperature-sensing thermister and an internal electrical heater for calibration so that, immersed in a constant-temperature bath, it could be used as a calorimeter. It was claimed that, when current was passed through the cell, deuterium fusion occurred within the cathode to yield low levels of neutrons, larger quantities of tritium, and large quantities of heat, many powers of 10 larger than would be expected from the neutron- and tritium-forming reactions. This last was the most striking claim and the basis for believing the cell might be an energy source.

As was evident to physicists or to anyone with an elementary knowledge of nuclear reactions, these claims were unexpected and astonishing. In order to overcome the electrostatic repulsion between deuterium nuclei, deuterium fusion should require enormous amounts of energy that would not normally be available at room temperature. Furthermore, the excited ^{4}He initially produced by fusion had always been observed to decompose to roughly equal quantities of tritium plus a proton and ^{3}He plus a neutron with the release of no additional energy beyond that coming from these products.

Some time later I told Chase Peterson the cautionary tale about the cigarette fiasco at Columbia, and perhaps I should have been warned, but Pons and Fleischmann were both scientists whom I knew and respected. My first inclination was to take their experimental results, particularly their calorimetry, which was closest to my own experience, at face value whatever the underlying phenomena involved. Because I was interested and as familiar as anyone with what they had been doing, and they and the University were being overwhelmed by inquiries and visitors, many of these were referred to me, and in this way I was drawn into the project. I reported on their claims at the spring meeting of the National Academy of Sciences and gave additional talks on the subject in subsequent weeks. After the National Cold Fusion Institute (NCFI) came into being during the summer, I became a consultant and advisor to it until it closed and I finally retired at the end of June 1991.

Almost immediately after the initial announcement I recognized that, if fusion was really occurring, some kind of electronic shielding would be required to lower internuclear repulsion. Furthermore, if the initial excited ^{4}He nucleus gave heat rather than neutrons and tritium, it must be giving ^{4}He by some process analogous to what is known in photochemistry as radiationless decay. I went to Jack Simons, our resident quantum mechanician, with the problem, and together we wrote a paper, soon published in the *Journal of Physical Chemistry* with the perhaps too accurate title "Two Innocent Chemists Look at Cold Fusion"[124]. In it Jack made a conventional quantum mechanical calculation of the shielding required, although we stated clearly that we had no model to suggest through which such shielding could arise. We further proposed some arguments why, if fusion did occur at these low energies, scission of excited ^{4}He to tritium rather than to neutrons might be favored, and that its lifetime might be long enough for decay to ground-state ^{4}He by some sort of coupling to the palladium lattice. While the paper was in press, Pons informed us that they had detected ^{4}He in the off-gases from a heat-producing cell, and we reported this in a footnote. Unfortunately, this observation was never confirmed at Utah. The paper has certainly been criticized, but I am not particularly embarrassed by it because we regarded it more as a statement of the problems that would have to be solved if the phenomenon was verified than a theoretical justification of the reported observations.

The weeks after the March 23 announcement were a time of high excitement at Utah. Cold fusion was getting worldwide attention. Although the average amount of "excess heat" reported in their paper was less than half of the heat input to the cell due to the electrolyis current, Fleischmann and Pons were now claiming "heat bursts" more than 10 times the latter. Apparent confirmations were coming in from other laboratories. On April 12 Pons gave a special invited lecture at the American Chemical Society meeting in Dallas to an audience of some 7000. The Utah Legislature voted $5,000,000 for cold fusion research, money that subsequently led to the setting up of the NCFI. On April 26, Fleischmann and Pons together with a group of university administrators, including Chase Peterson, appeared before a Congressional committee to stress the importance of their discovery and the need for federal support.

This was probably the high-water mark for cold fusion's popularity. A session on the topic at the annual meeting of the American Physical Society (which Fleischmann and Pons declined to attend) in late April was almost entirely critical and

Jane and myself winding up a tour of the Lesser Sunda Islands in Indonesia that we made with Hobart and Louise Young in 1989. My ceremonial hat was presented to me by a village chief on the island of Rota. (I think they have small heads!)

essentially voted that the phenomenon did not exist. There was also considerable criticism and questioning at the Electrochemical Society meeting in Los Angeles in early May at which Fleischmann and Pons spoke. Pons evidently found hostile debate very difficult to face. He stopped attending public meetings in the United States, and became increasingly unavailable even at Utah, where he appeared in the laboratory only in the mornings. His research group supervised the experiments but had no way of knowing how they were going because the data went on computer disks that Pons took home for analysis. Both Fleischmann and Pons became increasingly secretive and began to consider critics as enemies with whom they refused to deal, a disastrous course and a major source of the polarization between believers and critics, which still characterizes the subject.

At the NAS meeting my talk had been received courteously enough, but there was clearly extensive skepticism and some resentment over the manner of the March 23 announcement. I reported back to Fleischmann and Pons that there was serious criticism of the lack of controls and evidence for adequate mixing and homogeneity in their cells. If I was to talk about their work, I needed to know how reliable the calorimetry was. Fleischmann and Pons were collecting additional data with an automated data-collecting setup and very much wrapped up in what they called their "black box" model: setting up the differential equation for the complete behavior of their cells as calorimeters, numerically integrating, and doing a simultaneous least-squares fit to several parameters. I thought their approach was cumbersome and would be difficult to understand so I worked out a simpler one of my own. I needed data to compare different methods of analysis and the behavior of controls, but I was continually put off with the story that it would all be in the manuscript of an almost completed "definitive paper".

Finally, in September, I found two Fleischmann and Pons cells at NCFI said to be running for demonstration purposes although they were not giving excess heat. With the blessing of Hugo Rossi, a mathematician and former dean of the College of Science who was serving as the interim director of NCFI, I began some data collecting of my own. When Fleischmann and Pons learned of this, they were furious, demanded that the work stop, and finally stopped it by removing some of the equipment. I myself was now clearly one of the enemy and from there on got little information from either of them.

Before this happened I had been able to run a number of experiments over a range of current densities and convince myself that the cells gave heat balances reliable to at least 2–3%. None of these experiments indicated any significant excess heat, but just before the experiments were terminated I went to a higher current density. Here the computer data for one cell showed a sudden temperature rise of 4 °C that persisted for 2 hours and corresponded to at least 25% excess heat. Whatever the cause, these things do happen.

The rest of the history of the NCFI was largely one of increasing frustration and a running battle with Fleischmann and Pons. In October Rossi resigned as director because of disagreement with Fleischmann and Pons and what he considered inadequate backing by the administration. On February 1, 1990, Fritz Will, an electrochemist who had been a senior scientist at the General Electric Company, became director (James Brophy, the University Vice President for Research, held the position in the meantime). Will had been warned of some of the difficulties he might encounter and made repeated efforts to get Fleischmann and Pons's cooperation and to find out what they were doing, but with little success. By fall of 1990 he too was regarded as an enemy, Fleischmann and Pons were in Europe, and their laboratory at Utah was, as far as anyone could tell, closed down. I had become convinced, and so reported to Will and the directors of NCFI, that all the evidence suggested that Fleischmann and Pons had been having extraordinary difficulty and little success in reproducing their early results, particularly their flamboyant claims of large heat excursions. The elaborate charade of secrecy, unfulfilled promises to supply data, and threats of legal action to forestall criticism of their work had been an increasingly desperate attempt to keep anyone at Utah and the rest of the world from finding this out. As I stated at a meeting, they were in the position of a pair of prospectors who had picked up a piece of gold ore, organized a large mining company, and were attempting to keep from the stockholders the fact that they could not find the vein from which the ore had come. This attempt to put matters in local context was, I'm afraid, not very well received.

Up to now, the small amount that Fleischmann and Pons have published has added little to their initial announcement and seems to me to have been presented so selectively that it is almost impossible to evaluate without access to their data as a whole.

Outside of anything done by Fleischmann and Pons, the NCFI accomplished some reasonable scientific work that was written up in their final report—some of it was published. As examples, a group headed by Milton Wadsworth, a metallurgist and dean of the College of Mines, ran a number of cells of their own design with careful calibration and controls. Although they observed a few examples of apparent modest excess heat, reproducibility was so poor that the project was abandoned. Another group from the School of Engineering investigated a variety of designs for doing calorimetry on electrolytic cells ending with a simple, accurate, relatively inexpensive model containing a closed cell in which O_2 and D_2 were recombined, thus avoiding some of the difficulties with Fleischmann and Pons type open cells. In connection with this model, Andy Riley and I devised a simple method of monitoring the amount of deuterium absorbed by a Pd cathode during an electrolysis,[125] presumably an important parameter in connection with heat evolution, although it had not usually been measured. Unfortunately, the NCFI had been set up in a grandiose manner with extensive facilities and a large overhead on the assumptions that cold fusion was a reproducible phenomenon that would soon be verified and that money from industry and other sources would quickly flow in. When none of this happened it was in financial trouble, and Will, when he became director, found himself in command of a sinking ship. In spite of his efforts most projects never got beyond promising beginnings, and Utah's contribution to understanding what phenomena really underlay Fleischmann and Pons's original claims was, to me, very disappointing.

The broader story of cold fusion has spawned numerous articles and at least four books, the latest and most comprehensive being *Bad Science* by Gary Taubes.[125] Written from the point of view that the whole affair was a delusion tinged with fraud, it is heavily slanted in its presentation, although the facts it does report are for the most part quite accurate. The March 23 announcement was a terrible mistake, both in being premature and being too full of hype. Its timing seems to have been primarily to forestall any claims of priority by Stephen Jones and his colleagues at BYU. As such, it was both unnecessary and an act of bad faith because Jones and Fleischmann and Pons had previously agreed to simultaneous publication. Its hype gave the impression that here was something anyone could do with simple equipment and led to an enormous waste of time and money in many laboratories. It also led Peterson and his fellow enthusiasts to set up the NCFI to make Utah a leader in an as-yet hypothetical field, apparently over the advice of Fleischmann and Pons who sensed the shaky ground on which they were standing. The hype offended many scientists, particularly physicists who considered the idea preposterous as well as an invasion of their turf, but it provided a brief field day for the news media and a chance for the public to see some of scientists' foibles, which they might prefer to conceal. Aside from the waste of time and money, the chief losers were Fleischmann and Pons, who stirred up a storm they couldn't handle, and Chase Peterson, who resigned his presidency in 1991. As far as I can tell the University of Utah has suffered little long-term damage.

Where does this leave excess heat evolution, the most dramatic feature of cold fusion? To my view, apparent anomalous heat evolution during D_2O electrolysis at a Pd cathode has been reported by enough investigators with a large enough variety of calorimeters so that it seems to be a real phenomenon, although its occurrence is still quite unpredictable. Because it is so far a rare phenomenon, too few controls have been run to say unequivocally that it does not occur in ordinary water or with other metal cathodes. Arguments that it involves a nuclear process have so far been based on the fact that there is no other explanation and are much less convincing. Conclusive demonstration requires the detection of products that are obviously the consequence of nuclear reactions in quantities corresponding to the excess heat. At this writing there is a semiquantitative claim for the detection of ^4He in roughly the right quantity in the evolved gases from one set of experiments (implying a surface rather than a bulk process as originally claimed), but the measurement needs to be refined and confirmed in other laboratories if it is ever to convince most of the scientific public. Currently the field is in such bad repute in the United States that only a rash young investigator or student would touch it, and almost no money is available. I'd like to know the answer even if it does turn out to be trivial.

An Overview of Free Radical Chemistry

My story so far has been essentially a personal account of the adventures of my colleagues, my students, and myself in free radical chemistry over some 50 years. During this time, the field has grown from very modest beginnings to become a major part of organic chemistry. Thus it is now not only a body of knowledge in itself but also impinges widely on many other areas of chemistry.

Synthetic chemists are using radical reactions with increasing skill and effectiveness in complex synthesis. Radical intermediates are important in many technical processes, notably polymerizations and autoxidations, and are recognized as unwelcome participants in the aging and deterioration of organic materials. Radicals play a major part in atmospheric chemistry. Their role in biological processes, both in normal metabolism and in unwanted phenomena such as radiation damage and aging, although still far from completely understood, is developing a voluminous literature. Clearly my associates and I have played only a very small part in this growth, which has involved the efforts of hundreds of investigators all over the world. All I have been able to do in this account is to acknowledge a few of the ideas and results that particularly influenced our own work.

In closing my exposition, I will try to give an overview of this development in terms of the concepts involved and the sequence of their appearance. I would emphasize that many concepts appeared very early and that the accelerating growth of radical chemistry has been the result of their appreciation by a larger and larger fraction of the chemistry community and the development of new techniques and instruments that have greatly facilitated research.

The first step in the development of free radical chemistry, as I think of it, was simply the recognition that free radicals could be intermediates in chemical reactions. Initially, certainly, the most significant class of such processes were chain reactions, both because they were easily recognized by their susceptibility to initiation and inhibition and because of their practical importance. Radical sources are relatively rare, but in chain reactions, a single radical can convert many molecules of an easily available starting material into desirable products. Although such processes had been known to some physical chemists, their appreciation by organic chemists as applying to typical liquid-phase reactions first began in the late 1930s.

Once a chain reaction was recognized, a unique set of chain propagation steps could usually be deduced, or if different formulations were possible, a choice could be made on grounds of energetics, again an idea from older, gas-phase physical chemistry. As more and more chain processes were recognized, the major types of elementary steps could be distinguished, and in fact, the most important steps, radical displacements and additions, had been recognized in earlier gas-phase work. With a repertory of plausible elementary steps, chemists could now design new chain reactions for specific purposes. They have been doing so since the 1940s, but there has been a recent spectacular growth in this activity among synthetic chemists interested in constructing complex molecules.

Essential to this activity, and also to our understanding of radical processes, is a knowledge of structure–reactivity relations. Relative reactivities of different substrates toward a given radical can often be obtained easily from product studies on suitable competitive chain reactions. As we have seen, the first extensive body of data came from copolymerization and chain-transfer experiments in the 1940s. These data validated the method, and the principles involved, as well as the importance of energetics, steric hindrance, and polar effects, have proved to be very general. Many systems have been studied subsequently, and as I have pointed out, the ease of carrying out such studies has been enormously facilitated by new analytical methods such as gas chromatography.

The absolute rates of radical reactions (and the relative rates of reaction of different radicals) have presented a much more difficult problem. The best early data employed the rotating-sector technique, which was tedious and limited to reactions with very clean kinetics of the appropriate type. In recent years, new instrumental methods involving fast spectroscopy on the millisecond to nanosecond time scale have taken over. Rate constants for many elementary radical reactions have been measured, often together with their Arrhenius parameters. These, together with relative-rate measurements, have provided enough data so that the rate constants for most important elementary steps in radical reactions can now be at least reasonably estimated.

The demonstration that radicals are transient intermediates in nonchain processes has been more difficult, and the development, more piecemeal. Hey and Waters's[9] 1937 deduction that aryl radicals were formed in the decomposition of aryl diazonium derivatives and aryl peroxides was essentially by elimination of other possibilities, because the products did not correspond to those expected for electrophilic substitutions. As radical reactions became better characterized, this approach could be supplemented by showing parallels between products and those of known radical processes, by trapping intermediates with known radical traps, and by observing the initiation of radical chain processes. New instrumental methods have played an important role too: direct radical detection by electron spin resonance; CIDNP observation in suitable cases; and, for photoinduced processes, direct observation of radicals by fast spectroscopy.

At present, an investigator interested in radical reactions has an impressive armory of techniques and an enormous background of available information. Many of the main areas that have been long studied are well understood, although the design of new reaction sequences for synthetic purposes is certainly flourishing. Less well understood are the roles of radicals in oxidation–reduction reactions and in organometallic chemistry, the chemistry of radical ions, and the mysterious borderline between radical and ionic processes. I have discussed some of our own work in these areas.

The real frontier, though, in my view, is the understanding of the place of radicals as transient intermediates in living processes. For this area, we have chiefly tantalizing leads. Radiation damage to living organisms seems to be largely hydroxyl radical chemistry, and radicals have been proposed to play a part in aging processes. Some enzymatic processes, for example, those involving coenzyme B_{12}, look like radical processes. The list goes on but is largely conjectural. The detailed knowledge we now have about free radicals in simpler systems should make it possible for us to understand these phenomena in more detail, and here, I think, lies the most exciting developments for the future.

Ethics in Science

The Editor asked me to write something on my views of ethics in science. I think the matter has recently gotten more attention and generated more excitement than it deserves. Fraud and unethical behavior have always existed, but I think that they are rarer in science than in most human endeavors and that there is little evidence that they are increasing. Further, there is a tendency to include some items under the heading of unethical behavior that are simply reflections of bad judgment.

Ethics in science are not that different from ethics in other human activities. Primarily one should not lie, steal, or injure one's neighbors. This sounds a bit self-evident, but it is true, although the emphasis may be a bit different in the scientific world.

The cardinal sin in science is the kind of lying that involves fabricating or falsifying data because the validation of scientific hypotheses depends upon them. Fortunately, the system is to some extent self-healing. If the data are important, others will try to use or repeat them and their lack of validity will in all likelihood become apparent. Their net effect is thus the same as erroneous data reported in good faith. Unfortunately, the effort required to get a wrong result expunged from the scientific literature is usually much greater than would have been required to do the work right in the first place. Falsification is taken very seriously in science and can well end an investigator's career. This together with the strong probability of eventually being found out are major deterrents, and, I suspect, a chief reason why deliberate falsification is so rare.

Not only do falsified data and simply erroneous data have much the same effect, but the border between them is fuzzy and sometimes hard to determine. One sees this when confronting the question of what data to report and what to discard. In arguing for a hypothesis, there is a temptation to discard contradictory results or push them off as something to be looked into further later. On the other hand, there certainly are bad data that should be discarded. Too much "tailoring" of results to fit a particular hypothesis or to make a writer look good certainly borders on the unethical, but it is often necessary to make some selection if the results are to make any sense at all.

The pressure to publish and the race for priority produces another aspect of this dichotomy. For how long are results too sparse and sketchy to justify reporting them? Is acting too soon unethical, or simply bad judgment? Usually I'd vote for the latter, noting that editors and referees play an important role in answering the question. In both these cases it is important for an author to tell readers exactly what has and what has not been done so that they can judge for themselves the validity of the data and to avoid any appearance of impropriety. Scientists recognize that some of the published literature must be wrong, but they demand that the work in it be reported with due care and in good faith.

Stealing, in science, generally involves intellectual property. It appears as plagiarism of written material, appropriating another's ideas without giving them credit, or making improper use of "privileged" information such as one might obtain when acting as a referee. Because priority and reputation are so very important to scientists, any such appropriation of another's work is taken very seriously and it is wise to lean over backwards to avoid any appearance of improper behavior.

The last item, not harming your neighbor, conjures up the vision of the mad scientist working on devices for mass destruction, but this is really fiction. We're currently seeing enormous debate on what might or might not be potentially harmful and what public policies should be followed on questions in which scientific knowledge has some role. Except that their knowledge may make their opinions of particular value, scientists are in the same position as any other citizens. If you disapprove of something, don't work on it, and, if you think a dangerous mistake is being made, make your opinion known. Because there are sometimes real risks in doing this, everyone has to make their own decision on the proper course to follow.

A Final Word

While I was at Columbia, Jane and our children spent much of each summer in Jaffrey, New Hampshire, and I spent what time I could there by myself. In 1964 we bought a summer place there on a pond (in any place but New England it would be called a lake). We enlarged and winterized the house in stages, and, when I retired in the summer of 1991, becoming a professor emeritus, we moved to Jaffrey permanently with occasional escapes during the snowy winter months. Although I can no longer keep up with the literature and lack a convenient library I still do some writing and refereeing. I manage visits back to Utah and some other traveling (in 1992 I managed to see both North

Jane and I enjoying retirement in Jaffrey, New Hampshire, in the fall of 1991.

Cape and Cape Horn) and attend occasional scientific meetings. We're nearer to most of our children, and I think the move was sensible before I became completely redundant.

If I try to summarize my feelings about my career in a few words, I can't do better than to repeat what a number of scientists have said in the past: we've all been very fortunate. To someone of my temperament, being able to pursue my curiosity through research and reading with freedom and adequate support and to receive some recognition for my work have been all I could ask for. I can't think of anything I'd rather have done. The edifice of chemistry is built on a vast number of very small bricks, some load-bearing, some largely decorative, and a few corner- and keystones here and there. It is a satisfaction to have put in some of the bricks.

The structure has grown impressively in 50 years, and the view from the top gets better all the time. Besides, chemists make a congenial group, adept at sharing their enthusiasms. The international nature of science provides a marvelous excuse to travel, and I can take pleasure in a network of scientific friends all over the globe. I firmly believe that scientific work is basically useful and a social good. Although we may make dreadful misapplications of our knowledge along the way, the ability to understand the world is man's unique quality, and this understanding, to me, is his unique delight. A large share of this understanding lies in science and in chemistry, and "fortunate" is the right word.

I'm also conscious of the part that chance (or luck) has played in my career. My choice of working with Kharasch at Chicago, the circumstances that led to my going to U.S. Rubber with Frank Mayo, and the opening up of a position at Columbia were all turning points in my career. All of these occurred largely by chance, and when I decided to go into chemistry, I certainly didn't foresee the enormous growth of the field that was to carry me along. Some things that didn't happen might have brought more luck, but any number of events could have been less fortuitous.

In the almost 70 years since I first learned of it, chemistry and my own field of organic chemistry have seen enormous changes. From Kekulé to World War II was the age of "classical" or largely empirical organic synthesis from which molecular structures, including stereochemistry, could be deduced and on which it was possible to build an impressive chemical industry. From the 1940s to the 1960s, physical organic chemistry was dominant as the underlying mechanisms of organic reactions were worked out. This situation permitted a renaissance and enormous increase in the power of synthesis in the hands of chemists like R. B. Woodward and his successors, which continues to the present. At the same time there has been an almost unbelievable increase in the power of instrumental techniques, such as separation methods, NMR, exquisitely sensitive analytical methods, and the many applications of computers. These developments, in turn, make it increasingly possible to apply what we have learned to the chemistry of living organisms, and this application, I think, will be the major path of the future. Organic chemistry is not becoming just a service science as some have suggested; it is as central as ever, but is changing its direction.

On the other hand, chemical industry has certainly been facing difficulties. The flood of new products and processes has abated, and we have come to realize that procedures and actions that are innocuous on a small scale and for short times can be devastating if overdone. To me, there have been too many groups with a vested interest in promoting public hysteria, and there is no need to always affix "toxic" to the word "chemical". I hope we are working our way towards a more balanced outlook where it will be possible to again, if cautiously, speak of "better things for better living". All in all, I am optimistic and have many hopes for the future. One hope is that the satisfaction of careers in science, such as I have enjoyed, will continue to be open to young men and women and that many will choose this path for their lives.

References

1. Slosson, E. E. *Creative Chemistry;* Century: New York, 1919.
2. Fisher, C. H.; Walling, C. *J. Am. Chem. Soc.* **1939,** *61,* 1562, 1559.
3. Gomberg, M. *J. Am. Chem. Soc.* **1900,** *22,* 757.
4. Rice, F. O.; Rice, K. K. *The Aliphatic Free Radicals;* Johns Hopkins: Baltimore, MD, 1935.
5. Staudinger, H. *Ber. Dtsch. Chem. Ges.* **1920,** *53,* 1073.
6. Michaelis, L. *Chem. Rev.* **1935,** *16,* 243.
7. Bäckström, H. L. J. *Z. Phys. Chem. Abt. B* **1934,** *25,* 99.
8. Haber, F.; Willstätter, R. *Ber. Dtsch. Chem. Ges.* **1931,** *64,* 2844.
9. Hey, D. H.; Waters, W. A. *Chem. Rev.* **1937,** *21,* 169. I have recently reviewed the background of this review in more detail. Walling, C. *Chem. Br.* **1987,** *23,* 767.
10. Kharasch, M. S.; Engelmann, H.; Mayo, F. R. *J. Org. Chem.* **1937,** *2,* 288.
11. Flory, P. J. *J. Am. Chem. Soc.* **1937,** *59,* 241.
12. Kharasch, M. S.; Marker, R. *J. Am. Chem. Soc.* **1926,** *48,* 3130.
13. Kharasch, M. S.; Darkis, F. R. *Chem. Rev.* **1928,** *5,* 571.
14. Kharasch, M. S.; Mayo, F. R. *J. Am. Chem. Soc.* **1933,** *55,* 2468.
15. Mayo, F. R. *J. Chem. Educ.* **1986,** *63,* 97.
16. Kharasch, M. S.; Walling, C.; Mayo, F. R. *J. Am. Chem. Soc.* **1939,** *61,* 1559.
17. Walling, C.; Kharasch, M. S.; Mayo, F. R. *J. Am. Chem. Soc.* **1939,** *61,* 1711.

18. Goering, H. L.; Larsen, D. W. *J. Am. Chem. Soc.* **1959,** *81,* 5937.

19. Kharasch, M. S.; Brown, H. C. *J. Am. Chem. Soc.* **1939,** *61,* 2142.

20. Mayo, F. R.; Walling, C. *Chem. Rev.* **1940,** *27,* 351.

21. Mayo, F. R. *J. Am. Chem. Soc.* **1943,** *65,* 2324.

22. Mayo, F. R.; Lewis, F. M. *J. Am. Chem. Soc.* **1944,** *66,* 1594.

23. For chain transfer, *see* Hulbert, H. M.; Harman, R. A.; Tobolsky, A. V.; Erying, H. *Ann. N.Y. Acad. Sci.* **1943,** *12,* 205, 322.

24. Flory, P. J. *J. Am. Chem. Soc.* **1941,** *63,* 3083.

25. Walling, C. *J. Am. Chem. Soc.* **1945,** *67,* 441.

26. Walling, C.; Briggs, E. R.; Cummings, W.; Mayo, F. R. *J. Am. Chem. Soc.* **1950,** *72,* 48.

27. Alfrey, J., Jr.; Price, C. C. *J. Polym. Sci.* **1947,** 2, 101.

28. Hammett, L. P. *Physical Organic Chemistry;* McGraw-Hill: New York, 1941.

29. Hammett, L. P. *J. Am. Chem. Soc.* **1937,** *59,* 96.

30. Walling, C.; Briggs, E. R.; Wolfstirn, K. B.; Mayo, F. R. *J. Am. Chem. Soc.* **1948,** *70,* 1537.

31. Walling, C.; Seymour, D.; Wolfstirn, K. B. *J. Am. Chem. Soc.* **1948,** *70,* 1544.

32. Walling, C.; Seymour, D.; Wolfstirn, K. B. *J. Am. Chem. Soc.* **1948,** *70,* 2559.

33. Brown, H. C.; Okamoto, Y. *J. Am. Chem. Soc.* **1958,** *80,* 4979.

34. Bartlett, P. D.; Nozaki, K. *J. Am. Chem. Soc.* **1947,** *69,* 2299.

35. Walling, C.; Mayo, F. R. *Discuss. Faraday Soc.* **1947,** 2, 295.

36. Mayo, F. R.; Walling, C. *Chem. Rev.* **1950,** *46,* 191.

37. Bartlett, P. D.; Altschul, R. *J. Am. Chem. Soc.* **1945,** *67,* 816.

38. Bartlett, P. D.; Swain, C. G. *J. Am. Chem. Soc.* **1945,** *67,* 2273.

39. Smith, W. V.; Ewart, R. W. *J. Chem. Phys.* **1948,** *16,* 592.

40. Ingold, K. U.; Bowry, V. W.; Stocker, R.; Walling, C. *Proc. Natl. Acad. Sci. U.S.A.* **1993,** *90,* 45.

41. A detailed account of this impressive achievement is given in *Synthetic Rubber;* Witby, G. S., Ed.; Wiley: New York, 1954.

42. Kharasch, M. S.; Jensen, E. V.; Urry, W. H. *Science (Washington, D.C.)* **1945,** *102,* 128.

43. A summary is given by Walling, C.; Huyser, E. S. In *Organic Reactions;* Cope, A. C., Ed.; Wiley: New York, 1963; Vol. 13.

44. Waters, W. A. *The Chemistry of Free Radicals;* Oxford University: London, 1946.

45. For a review of British work, *see* Bateman, L. *Q. Rev. Chem. Soc.* **1954,** *8,* 147.

46. Walling, C. *J. Am. Chem. Soc.* **1950,** *72,* 1164.

47. Walling, C.; Peisach, J. *J. Am. Chem. Soc.* **1958,** *80,* 5819.

48. Walling, C.; Schugar, H. J. *J. Am. Chem. Soc.* **1963,** *85,* 607.

49. Walling, C.; Naimen, M. *J. Am. Chem. Soc.* **1962,** *84,* 2628.

50. Ohtsuka, Y.; Walling, C. *J. Am. Chem. Soc.* **1966,** *88,* 4167.

51. Walling, C.; Pellon, J. *J. Am. Chem. Soc.* **1957,** *79,* 4776, 4782, 4786.

52. Walling, C. *J. Polym. Sci.* **1960,** *48,* 335.

53. *The Chemistry of Petroleum Hydrocarbons;* Brooks, B. J.; Boord, C. E.; Kurtz, S. S., Jr.; Schmerling, L., Eds.; Reinhold: New York, 1955; Vol. 3.

54. Walling, C. *Free Radicals in Solution;* Wiley: New York, 1957.

55. Walling, C.; Chang, Y.-W. *J. Am. Chem. Soc.* **1954,** *76,* 4878.

56. Walling, C.; Heaton, L. *J. Am. Chem. Soc.* **1965,** *87,* 38.

57. Walling, C.; Buckler, S. A. *J. Am. Chem. Soc.* **1953,** *75,* 4372.

58. Walling, C.; Miller, B. *J. Am. Chem. Soc.* **1957,** *79,* 4181.

59. Eibner, A. *Ber. Dtsch. Chem. Ges.* **1903,** *36,* 1229.

60. Miller, B.; Walling, C. *J. Am. Chem. Soc.* **1957,** *79,* 4187.

61. Russell, G. A. *J. Am. Chem. Soc.* **1957,** *79,* 4181.

62. Walling, C.; Mayahi, M. F. *J. Am. Chem. Soc.* **1959,** *81,* 1485.

63. Bunce, N. J.; Ingold, K. U.; Landers, J. P.; Lusztyk, J.; Scaiano, J. C. *J. Am. Chem. Soc.* **1985,** *107,* 5464.

64. Bloomfield, G. F. *J. Chem. Soc.* **1944,** 114.

65. Walling, C.; Jacknow, B. B. *J. Am. Chem. Soc.* **1960,** *82,* 6108.

66. Walling, C.; Thaler, W. *J. Am. Chem. Soc.* **1961,** *83,* 3877.

67. Walling, C.; Wagner, P. J. *J. Am. Chem. Soc.* **1964,** *86,* 3368.

68. Walling, C.; Padwa, A. *J. Am. Chem. Soc.* **1963,** *85,* 1597.

69. Adam, J.; Gosselain, P. A.; Goldfinger, P. *Nature (London)* **1953,** *171,* 704.

70. Walling, C.; Rieger, A. L.; Tanner, D. D. *J. Am. Chem. Soc.* **1963,** *85,* 3129.

71. Russell, G. A.; Desmond, K. M. *J. Am. Chem. Soc.* **1963,** *85,* 3139.

72. Pearson, R. E.; Martin, J. C. *J. Am. Chem. Soc.* **1963,** *85,* 3142.

73. Walling, C.; McGuinness, J. A. *J. Am. Chem. Soc.* **1969,** *91,* 2053.

74. Walling, C.; Bristol, D. *J. Org. Chem.* **1972,** *37,* 3514.

75. Walling, C.; Clark, R. T. *J. Am. Chem. Soc.* **1974,** *96,* 4530.

76. Day, J. C.; Lindstrom, M. J.; Skell, P. S. *J. Am. Chem. Soc.* **1974,** *96,* 5616.

77. Tanner, D. D.; Reed, D. W.; Tan, S. L.; Meintzer, C. P.; Walling, C. *J. Am. Chem. Soc.* **1985,** *107,* 6576.

78. Luning, V.; Seshadri, S.; Skell, P. S. *J. Org. Chem.* **1986,** *51,* 2071.

79. Walling, C.; Helmreich, W. *J. Am. Chem. Soc.* **1959,** *81,* 1144.

80. Walling, C.; Rabinowitz, R. *J. Am. Chem. Soc.* **1959,** *81,* 1137.

81. Hoffman, F. W.; Ess, R. J.; Simmons, T. C.; Hanzel, R. S. *J. Am. Chem. Soc.* **1956,** *78,* 6414.

82. Ramirez, F.; McKelvie, N. *J. Am. Chem. Soc.* **1957,** *79,* 5829.

83. Walling, C.; Rabinowitz, R. *J. Am. Chem. Soc.* **1959,** *81,* 1243.

84. Walling, C.; Pearson, M. S. *J. Am. Chem. Soc.* **1964,** *86,* 2262.

85. A good review is given by Beckwith, A. L. J.; Ingold, K. U. In *Rearrangements in Ground and Excited States;* Academic: New York, 1980; Vol. 1.

86. Menapace, L. W.; Kuivila, H. G. *J. Am. Chem. Soc.* **1964,** *86,* 3047.

87. Walling, C.; Cooley, J. H.; Ponaras, A. A.; Racah, E. J. *J. Am. Chem. Soc.* **1966,** *88,* 5361.

88. Walling, C.; Cioffari, A. *J. Am. Chem. Soc.* **1972,** *94,* 6059, 6064.

89. Walling, C.; Indictor, N. *J. Am. Chem. Soc.* **1958,** *80,* 5814.

90. Walling, C.; Hodgdon, R. B., Jr. *J. Am. Chem. Soc.* **1958,** *80,* 228.

91. Walling, C.; Zhao, C. *Tetrahedron* **1982,** *38,* 1105.

92. Zhao, C.; El-Taliawi, G. M.; Walling, C. *J. Org. Chem.* **1983,** *48,* 4908.

93. Bartlett, P. D.; Leffler, J. E. *J. Am. Chem. Soc.* **1950,** *72,* 3030.

94. Walling, C.; Waits, H. P.; Milovanovic, J.; Papaioannou, C. G. *J. Am. Chem. Soc.* **1970,** *92,* 4927.

95. Walling, C.; Sloan, J. P. *J. Am. Chem. Soc.* **1979,** *101,* 7679.

96. Walling, C.; Humphreys, R. W. R.; Sloan, J. P.; Miller, T. *J. Org. Chem.* **1981,** *46,* 5261.

97. Taylor, K. Q.; Gorindan, C. K.; Kaelin, M. S. *J. Am. Chem. Soc.* **1979,** *101,* 2091.

98. Lawler, R. G.; Barbara, P. F.; Jacobs, D. *J. Am. Chem. Soc.* **1978,** *100,* 4912.

99. Luner, C.; Szwarc, M. *J. Chem. Phys.* **1955,** *23,* 1978.

100. Walling, C.; Gibian, M. J. *J. Am. Chem. Soc.* **1965,** *87,* 3413.

101. Walling, C.; Gibian, M. J. *J. Am. Chem. Soc.* **1965,** *87,* 3367.

102. Walling, C.; Naglieri, A. *J. Am. Chem. Soc.* **1960,** *82,* 1820.

103. Walling, C. *Acc. Chem. Res.* **1983,** *16,* 448.

104. *A Century of Chemistry;* Skolnik, H.; Reese, K. M., Eds.; American Chemical Society: Washington, DC, 1976; pp 65–72.

105. Breslow, R.; Corcoran, R. J.; Snider, B. B.; Doll, R. J.; Khanna, P. L.; Kaleya, R. *J. Am. Chem. Soc.* **1977,** *94,* 905.

106. Stork, G.; Mook, R., Jr. *J. Am. Chem. Soc.* **1987,** *108,* 2829.

107. Fraenkel, G. K., Hirshon, J. M.; Walling, C. *J. Am. Chem. Soc.* **1954,** *76,* 3606.

108. Gardner, D. M.; Fraenkel, G. K. *J. Am. Chem. Soc.* **1856,** *78,* 3479.

109. Walling, C.; Stacey, F. R.; Jamison, S. E.; Huyser, E. S. *J. Am. Chem. Soc.* **1958,** *80,* 4543, 4546.

110. Walling, C.; Brown, E.; Bartz, K. W. U.S. Patent 3,027,352, March 27, 1962.

111. Walling, C.; Lepley, A. R. *J. Am. Chem. Soc.* **1972,** *94,* 2007.

112. Schulman, E. M.; Bertrand, R. D.; Grant, D. M.; Lepley, A. R.; Walling, C. *J. Am. Chem. Soc.* **1972,** *94,* 5972.

113. Merz, J. H.; Waters, W. A. *J. Chem. Soc.* **1949,** S15.

114. Walling, C.; Kato, S. *J. Am. Chem. Soc.* **1971,** *93,* 4275.

115. Walling, C. *Acc. Chem. Res.* **1975,** *8,* 125.

116. Walling, C.; Goosen, A. *J. Am. Chem. Soc.* **1973,** *95,* 2987.

117. Walling, C.; Partch, R. E.; Weil, T. *Proc. Natl. Acad. Sci. U.S.A.* **1975,** *72,* 140.

118. Norman, R. O. C.; Storey, P. M. *J. Chem. Soc. B* **1970,** 1099.

119. Walling, C.; Camaioni, D.; Kim, S. S. *J. Am. Chem. Soc.* **1978,** *100,* 4814.

120. Walling, C.; El-Taliawi, G. M.; Amarnath, K. *J. Am. Chem. Soc.* **1984,** *106,* 7573.

121. Walling, C. *J. Am. Chem. Soc.* **1975,** *97,* 2A.

122. Fleischmann, M.; Pons, S. *J. Electroanal. Chem.* **1989,** *261,* 301.

123. Walling, C.; Simons, J. *J. Phys. Chem.* **1989,** *93,* 4693

124. Riley, A. M.; Seader, J. D.; Pershing, D. W.; Walling, C. *J. Electrochem. Soc.* **1992,** *139,* 1342 .

125. Taubes, G. *Bad Science*; Random House: New York, 1993.

Index

A

Academic research, Columbia University, 1952–1969, 45–81
Active centers, copolymer compositions, 29
Alkoxy radicals, production, 54
American Chemical Society (ACS)
 Committee on Professional Training (CPT), 69–71
 editor of JACS, 93–97
Amine reactions, production of radicals, 61–62
Anderson, R. T., research program in fundamental polymer chemistry, 22
Ando, W., stone lantern factory, Tsukuba, Japan, 1978 (photo), 99
Antigua Castor Enterprises, castor beans and hotel, 76
Aromatic cations
 cleavage of complex side chains, 92
 reversible hydration, 90
Autoxidation, Faraday Society discussion, 1947, 36–40
Azo dyes, DuPont, 20

B

Bailar, John, American Chemical Society meeting, 1972 (photo), 69
Bamford, C. H., polymer chemist, 39
Bartlett, Paul
 group seminars and bicycle trips, 43
 luncheon of physical organic chemists, Washington, D.C., 1983 (photo), 101
 National Science Foundation Summer Institute, Durango, Colorado, 1960 (photo), 79
 polymer seminars, 26
 Stanford Research Institute, 1968 (photo), 31
 summer course on physical organic chemistry, 79–81
Bateman, L., hydrocarbon autoxidation, 39–40
Baxter, G. P., quantitative analysis course, 8
Beckwith, Athel
 70th birthday symposium, Park City, Utah, 1986 (photo), 107
 Australian National University cheery greeting, 1989 (photo), 103

Beckwith, Athel—*Continued*
Australian travel arrangements, 102
Symposium on Organic Free Radicals, St. Andrews, Scotland, 1984 (photo), 106
Beckwith, Kaye, Symposium on Organic Free Radicals, St. Andrews, Scotland, 1984 (photo), 106
Bent, Henry, physical chemistry course, 8
Bentrude, Wes
70th birthday symposium, Park City, Utah, 1986 (photo), 107
free radical symposium in Zurich, 1988 (photo), 104
Bloomfield mechanism, NBS reactions, 56
Bond dissociation energy, calculation in radical reactions, 18
Bordwell, F. G., 3rd Reaction Mechanisms Conference, Northwestern University, 1950 (photo), 42
Bradley, General Omar, Harvard commencement, 1947 (photo), 39
Breslow, Ronald
radical relay technique, 72
senior organic staff at Columbia (photo), 67
Bridgeman, P. W., high-pressure technology, 46–47
Bromine bridging, two successive *trans* additions, 15
Brooklyn Poly, polymer institute, 26
Brophy, James
interim director of NCFI, 111
press conference on cold fusion, 108–109
Brown, H. C.
radical chain halogenations, 16
Reaction Mechanisms Conference, Northwestern University, 1950 (photo), 42
Buckler, Sheldon, autoxidation of Grignard reagents, 51
t-Butyl hypochlorite, chlorination at allylic positions, 54–55
Butyl rubber, wartime synthetic rubber program, 34
Butyraldehyde, polymerization, 47–48

C

Cairns, T. L., career, 19
Carbonium ions, "nonclassical", 68
Carboxyl inversion reaction, analogue, 66
Cattle boat, DuPont ferry, 19
Celanese Corporation, consultant, 73–75
Celcon, development, 74
Chain carrier, succinimide radical, 53
Chain reactions
propagation steps, 115–116
radical sources, 115
Chang, Y.-W., *t*-butyl hydroperoxide, 50–51
Charge-transfer concept, molecular complexes, 30–31
Chemical dermatitis, dioxins, 20
Chemical Warfare Service, research, 73
Chemically induced dynamic nuclear polarization (CIDNP), 86–87
Chemistry set, introduction to chemistry, 1
Chinese Academy of Sciences, travel to People's Republic of China, 1982, 102
Chlorinating agent, *t*-butyl hypochlorite, 53–54
Cigarette filter, effectiveness, 75
Cioffari, Angela, cyclopentane derivatives, 60
Cohen, Saul G., Harvard classmate, 8
Cold fusion
frustrating project, 103–114
results of early publicity, 113
Columbia University
academic research, 1952–1969, 45–81
academic scene, 66–68
high-pressure laboratory, 1955 (photo), 46
Committee on Professional Training, American Chemical Society, 69–71

Computer network, journal publishing, 97
Cooley, J. H., cyclopentane derivatives, 60
Criegee, R., intermediacy of hydroperoxides in autoxidations, 40
Cristol, Stan, Gordon Research Conference, New Hampton, N.H., 1971 (photo), 27

D

ΔH, calculation in radical reactions, 18
Daly, Reginald A. (uncle), emeritus professor of geology at Harvard, 46
Data, falsified and erroneous, 118
Dauben, W. G., synthetic utility of high pressures, 49
Dawson, Charles R.
 introduction to Columbia University, 66
 senior organic staff at Columbia (photo), 67
Debye, P. P.
 emulsion polymerization, 33
 polymer lectures, 27
Degradative chain transfer, inhibitor reactions, 32
Deuterium fusion, press conference, 108–109
Dewar, M. J. S., polymer chemist, 38–39
Diels–Alder reaction, synthesis of cantharidin, 47
Doak, Ken, farewell luncheon, U.S. Rubber Co. (photo), 25
DuPont
 employment, 1939–1943, 19–22
 financial support, 72–73
 safety procedures, 20
Dye Works, DuPont Jackson Laboratories, 19

E

Editorial selection, scientific journals, 96–97
Education
 Harvard University, 6–10
 introduction to chemistry, 1
 North Shore Country Day School, 3
 University of Chicago, 11–18
El Taliawi, Gamil
 cleavage of complex side chains, 92
 Egyptian nightlife, 1974 (photo), 98
 hydroxyl radicals, 88
 University of Cairo, 98
Electronic shielding, cold fusion, 105
Eliel, Ernest, luncheon of physical organic chemists, Washington, D.C., 1983 (photo), 101
Emulsion polymerization
 butyraldehyde, 47–48
 technique, 32–33
 wartime synthetic rubber program, 34
Ethics in science, 117–119
European tour, Nazi Germany, 11
Ewart, R. W.
 emulsion polymerization, 33
 research program in fundamental polymer chemistry, 22
Eyring, Henry, University of Utah, 1970 (photo), 83

F

Farmer, E. H., hydrocarbon autoxidation, 39–40
Fenton reagent, selectivities, 89
Fieser, Louis, introductory organic course, 6
Fisher, C. H. (Hap), side-chain reactions of benzene derivatives, 6
Fleischmann, Martin
 heat bursts, 108
 secretive behavior, 110–112
Fletcher, James, effective persuasion, 82
Foote, Chris, International Symposium on Free Radicals, Freiburg, Germany, 1981 (photo), 106
Fraenkel, George, free radicals in polymerizing methyl methacrylate, 72

Free radical additions, carbon–carbon bonds, 35–36
Free radical chemistry
other developments, 32–35
overview, 114–117
solutions, 49–50
study by physical chemists, 11–12
Free radical research, new lines, 50–57

G

Gale, Henry, University of Chicago, 11
Gel formation, intramolecular reactions in radical systems, 27–28
Gibbons, W. A., research program in fundamental polymer chemistry, 22
Gibian, Morton J., photosensitizers for peroxides, 65
Glassware, Welch Scientific Company, 5
Goldfinger mechanism, NBS reactions, 56
Gonikberg, M. G., polymer meeting in Moscow, 1960 (photo), 48
Goosen, Andre, HO•–substrate reactions, 89
Grant, David
associate editor, 93
chemically induced dynamic nuclear polarization, 87
effective persuation, 82
Greene, Fred, meeting of American Chemical Society editors, 1978 (photo), 94
Gregg, Bob, farewell luncheon, U.S. Rubber Co. (photo), 25
Grignard reagents, autoxidation, 51

H

Halogen carriers, hypochlorite chemistry, 53–57
Hammett, Louis
chairman at Columbia University, 45
introduction to Columbia University, 66
Hammett acidity function, silica–alumina and similar catalysts, 41–42
Hammett equation, application to radical reactions, 30
Hammond, George, summer course on physical organic chemistry, 79–81
Harkins, W. D., emulsion polymerization, 33
Hart, E. J., career, 41
Harvard University, undergraduate study, 6–10
Haskell, Frederika C. (mother), genealogy, 1–2
Hawthorne, Fred, with plane, Riverside, California, 1967 (photo), 81
Heat bursts, cold fusion, 108
Heat evolution, D_2O electrolysis at Pd cathode, 113
Heaton, LaDonne, *t*-butyl hydroperoxide, 50–51
Helmreich, Wolf, addition step reversibility, 57
High-pressure reactions, research program, 45–49
Hock, H., intermediacy of hydroperoxides in autoxidations, 40
Huisgen, Rolf, Gordon Research Conference, New Hampton, N.H., 1971 (photo), 27
Hutchins, Robert M., University of Chicago, 10
Huyser, Earl, Chemical Warfare Service, 73
Hydroxyl radical chemistry, long-lasting project, 87–89
Hypochlorite chemistry, halogen carriers, 53–57

I

Industry
E. I. du Pont de Nemours and Company, 1939–1943, 19–22
Lever Brothers Company, 1949–1952, 40–44

Industry —*Continued*
U.S. Rubber Company, 1943–1949, 22–40
Ingold, Cairine, 3rd International Symposium on Organic Free Radicals, St. Andrews, Scotland, 1984 (photo), 106
Ingold, K. U., laser flash photolysis, 52
Ingold, Keith
70th birthday symposium, Park City, Utah, 1986 (photo), 107
Free Radicals in Perspective symposium, 104
Reaction Mechanisms Conference, Northwestern University, 1950 (photo), 42, 43
Symposium on Organic Free Radicals, St. Andrews, Scotland, 1984 (photo), 106
Isomers, 2-pentene, 13–15

J

Jacknow, B. B., *t*-butoxy radicals, 53
Jackson Laboratories, DuPont, 19
Japanese Society for the Promotion of Science, six weeks in Japan, 98
Johnson, Richard, reaction of ethylene glycol with Fenton's reagent, 90
Jones, Stephen, nuclear reaction during electrolysis of D_2O, 104
Journal of the American Chemical Society (JACS)
archival and abstract forms, 97
editor, 93–97
editorial policy, 94–95
referees, 95–96
Julia, Marc, Park City, Utah, 1980 (photo), 59
Julia, Mme., Park City, Utah, 1980 (photo), 59

K

Kato, Shin'ichi, hydroxyl radicals, 88
Katz, Tom, Columbia University organic staff, 72
Ketones, triplet state, 65
Kharasch, Morris
2-pentene isomers, 13–15
Reaction Mechanisms Conference, Colby Jr. College (photo), 17
research group, University of Chicago, 15–16
University of Chicago, 11
King, Ed, Hallett Peak, Rocky Mountain National Park, 1965 (photo), 80
Kochi, Jay, 70th birthday symposium, Park City, Utah, 1986 (photo), 107
Kohler, E. P., advanced organic course, 7

L

Lamb, Arthur, JACS editor, 93
Lepley, A. R., chemically induced dynamic nuclear polarization, 87
Leslie, John H., Stanley Steamer (photo), 5
Lever Brothers Company
consultant, 73
employment, 1949–1952, 40–44
Lewis, Fred M.
career, 41
early work on copolymerization, 25
farewell luncheon, U.S. Rubber Co. (photo), 25
Little, J. R., ethylene polymerization, 36
Liu, You Cheng
free radical chemistry, 102
Lanzhou University, People's Republic of China, 1982 (photo), 100

M

Mark, Herman, polymer institute, 26
Marshall, General George, Harvard commencement, 1947 (photo), 39
Matheson, Max
career, 41
farewell luncheon, U.S. Rubber Co. (photo), 25

Mayo, Ellie
Oxford University, England, 1960 (photo), 78
Yosemite National Park, California, 1979 (photo), 105
Mayo, Frank
career, 41
Free Radicals in Perspective symposium, 104
Kharasch's research group, 16
Oxford University, England, 1960 (photo), 78
Reaction Mechanisms Conference, Colby Jr. College (photo), 17
research program in fundamental polymer chemistry, 22
Stanford Research Institute, 1968 (photo), 31
U.S. Rubber Company, 1947 (photo), 37
Yosemite National Park, California, 1979 (photo), 105
McGuinness, Jim, hypochlorite chlorinations, 56
Melville, H. W., polymerization kinetics, 38
Miller, Bernard, effect of ring substitution on radical side-chain chlorination of toluene, 51–52
Mintz, Mike, Park City, Utah, 1986 (photo), 54
Model airplanes, rocket-propelled models, 4
Morrison, Jim, Australian travel arrangements, 102

N

Naglieri, A., nitrogen analogues of benzoyl peroxide, 66
Nakanishi, Koji, Columbia University organic staff, 72
National Cold Fusion Institute (NCFI), establishment, 110
National Science Foundation (NSF)
financial support, 73
summer course on physical organic chemistry, 79–81
Nazi Germany, European tour, 11
Neuman, R. C.
70th birthday symposium, Park City, Utah, 1986 (photo), 107
pressure to determine transition-state volumes, 49
Newman, M., 3rd Reaction Mechanisms Conference, Northwestern University, 1950 (photo), 42
"Nonclassical" ions, carbocations, 68
Norbornyl system, peculiar geometry, 68
Norman, R. O. C., acid-catalyzed dehydration to radical cations, 90
Norrish, R. G. W., flash photolysis, 38
Nucleophilic attack on peroxide, production of radicals, 61
Nylon, metal–chelate dyes, 20

O

Oae, Shigaru
International Symposium on Free Radicals, Freiburg, Germany, 1981 (photo), 106
travel arrangements, 98
Office of Naval Research, financial support, 73
Office of Ordnance Research, financial support, 73
Office of Scientific Research and Development (OSRD), antimalarial drug program, 25–26
Oxidation–reduction reactions, free radical chemistry, 117
Oxidative degradation, rubber and other polymers, 40

P

Padwa, Albert
70th birthday symposium, Park City, Utah, 1986 (photo), 54, 107
β-scission pattern, 55–56
Parry, Bob
associate editor, 93
University of Utah, 1970 (photo), 83
Pearson, Myrna, radical cyclizations, 59

Pearson, R., 3rd Reaction Mechanisms Conference, Northwestern University, 1950 (photo), 42
Pedersen, C. J., DuPont research worker, 20
Peisach, Jack
high-pressure laboratory, 1955 (photo), 46
high-pressure work, 47
Pellon, Joseph
high-pressure laboratory, 1955 (photo), 46
high-pressure work, 47
2-Pentene, addition of hydrogen bromide, 13–15
Permanent wave, odorless, 41
Peroxide chemistry, research, 61–66
Peroxide decomposition
rates and mechanisms, 62–63
transition-state model, 63–65
Peroxide effect, bromopentane system, 14–15
Peterson, Chase
federal support for cold fusion, 108
press conference on cold fusion, 108–109
Petroleum Research Fund of the ACS, financial support, 73
Plagiarism in science, 119
Polar effect
polymerization reactions, 29
quantitative general theory, 31–32
Polyethylene, wartime development, 36
Polymer chemistry, U.S. Rubber Company, 1943–1949, 22–40
Polymerization initiation, *t*-butyl hydroperoxide, 50–51
Polymerization reactions
British view, 39
rate constants for chain propagation and termination, 32
Polytechnic Institute of New York, polymer institute, 26
Pons, Stanley
^{4}He in off-gases from heat-producing cell, 108
heat bursts, 108
Pons, Stanley—*Continued*
nuclear reaction during electrolysis of D_2O, 104
secretive behavior, 110–112
Porter, George, flash photolysis, 38
Porter, Ned, 70th birthday symposium, Park City, Utah, 1986 (photo), 107
Prelog, Vladimir, E.T.H., Zurich, 1960 (photo), 78
Pryor, Bill, spring ACS meeting, St. Louis, 1984 (photo), 52
Publications
Free Radicals in Solution, 49–50
peroxide effect, 16–18
race for priority, 118–119
side-chain reactions of benzene derivatives, 6

R

Rabinowitz, Bob
light-catalyzed radical reaction, 58
reversibility of displacement step, 57
Radical cations, oxidation intermediates, 90–93
Radical chemistry, structure and reactivity, 27–32
Radical cyclizations, chain reaction, 58–60
Radical pair–ion pair, peroxide decomposition, 63–65
Radical polymerization, alternating effect, 28–29
Ramirez, Fausto, light-catalyzed radical reaction, 58
Reaction rates, free radical chemistry, 116
Referees, invaluable service, 95–96
Research, Columbia University, 45–81
Resonance hybrids, polymerization reactions, 29–30
Retirement, Jaffrey, New Hampshire, 119
Rieger, Nancy, Park City, Utah, 1986 (photo), 54
Rieger, Phil, Park City, Utah, 1986 (photo), 54

Riley, Andy, deuterium absorption, 113
Ring substitution, effect on radical side-chain chlorination of toluene, 51–52
Roberts, Edith
Gordon Research Conference, New Hampton, N.H., 1971 (photo), 27
National Science Foundation Summer Institute, Durango, Colorado, 1960 (photo), 79
Roberts, Jack
geographic preferences, 45
Gordon Research Conference, New Hampton, N.H., 1971 (photo), 27
molecular orbital theory, 43
National Science Foundation Summer Institute, Durango, Colorado, 1960 (photo), 79
Reaction Mechanisms Conference, Northwestern University, 1950 (photo), 42
seminars and meetings, 41
summer course on physical organic chemistry, 79–81
Robinson, Sir Robert, British organic chemistry establishment, 38
Rossi, Hugo, interim director of NCFI, 111
Ruppert, Phil, castor beans in Antigua, 76
Russell, Glen
alkane chlorinations, 52
spring ACS meeting, St. Louis, 1984 (photo), 52
Ruzicka, Leopold, E.T.H., Zurich, 1960 (photo), 78

S

Sailing, camp on Cape Cod, 4
Sakurai, Hideki
Salt Lake City, 1979 (photo), 100
travel arrangements, 98
Sauer, Charlotte
editorial secretary, 93
lab party, University of Utah, 1989 (photo), 95
Scaiano, Tito, free radical symposium in Zurich, 1988 (photo), 104
Schulman, Dianne, Salt Lake City, 1973 (photo), 87
Schulman, Ed
chemically induced dynamic nuclear polarization, 87
Salt Lake City, 1973 (photo), 87
Seeman, Jeff, luncheon of physical organic chemists, Washington, D.C., 1983 (photo), 101
Simons, Jack, electronic shielding for cold fusion, 105
Single electron transfer (SET)
polar effect, 32
production of radicals, 61
Skell, P. S., NMS reactions, 56
Smith, P. A. S., 3rd Reaction Mechanisms Conference, Northwestern University, 1950 (photo), 42
Smith, W. V., emulsion polymerization, 33
Smith–Ewart theory, rate of emulsion polymerization, 33
Smoke generation, appallingly successful experiment, 6
Stacey, Frank, Chemical Warfare Service, 73
Stanley Steamer, steam engineering and car repair, 4–5
Steric hindrance, radical additions, 28
Stockmayer, W. H., seminars and meetings, 41
Stork, Gilbert
70th birthday symposium, Park City, Utah, 1986 (photo), 107
Columbia University organic staff, 72
radical cyclizations, 72
Story, Gilbert, synthesis of cantharidin, 47
Structure–reactivity relations, free radical chemistry, 116
Swain, C. G., seminars and meetings, 41
Synthetic detergents, Lever Brothers Company, 40–44

T

Tanner, Dennis, 70th birthday symposium, Park City, Utah, 1986 (photo), 107
Tarbell, Stan, luncheon of physical organic chemists, Washington, D.C., 1983 (photo), 101
Thiol additions, chain steps, 57
Tishler, Max, laboratory instructor, 7–8
Toluene, effect of ring substitution on radical side-chain chlorination, 51–52
Transient intermediates
living processes, 117
nonchain processes, 116–117
Transient radicals, study by physical chemists, 11–12
Travel
Australia, 1987, 102–103
Egypt, 1974, 98
European tour, 1960, 49
Faraday Society discussion, 1947, 36–40
Japan, 1978, 98
People's Republic of China, 1982, 102
sabbatical leave, 1960, 79–81
University of Washington, Seattle, 1958, 77
Triplet energy transfer, ionic products, 65–66
Trozzolo, Tony, luncheon of physical organic chemists, Washington, D.C., 1983 (photo), 101
Turro, Nick
Columbia University organic staff, 72
senior organic staff at Columbia University (photo), 67

U

University of Chicago, graduate school, 13–18
University of Colorado, Boulder, summer course, 1965, 81
University of Utah
1969 to retirement, 82–120
academic scene, 84
geographical setting, 85–86
Unzipping, polyacetal chain, 74
U.S. Rubber Company, employment, 1943–1949, 22–40

V

Vinyl polymerization, basic tenets, 23–24

W

Wadsworth, Milton, modest excess heat, 112
Wagner, Peter
70th birthday symposium, Park City, Utah, 1986 (photo), 54, 107
solvent effects, 55
spring ACS meeting, St. Louis, 1984 (photo), 52
Walling, Barbara (daughter)
70th birthday symposium, Park City, Utah, 1986 (photo), 107
Fairbanks, Alaska, 1967 (photo), 70
Walling, Cheves
1946 family group (photo), 21
70th birthday symposium, Park City, Utah, 1986 (photo), 107
age five (photo), 2
American Chemical Society meeting, 1972 (photo), 69
Egyptian nightlife, 1974 (photo), 98
farewell luncheon, U.S. Rubber Co. (photo), 25
French Lick, Indiana, vacation (photo), 9
Gordon Research Conference, New Hampton, N.H., 1971 (photo), 27
Great Wall of China, 1982 (photo), 101
Hallett Peak, Rocky Mountain National Park, 1965 (photo), 80
Harvard (photo), 7
high-pressure laboratory, 1955 (photo), 46

Walling, Cheves—*Continued*
International Symposium on Free Radicals, Freiburg, Germany, 1981 (photo), 106
lab party, University of Utah, 1989 (photo), 95
Lanzhou University, People's Republic of China, 1982 (photo), 100
Lesser Sunda Islands, Indonesia, 1989 (photo), 110
luncheon of physical organic chemists, Washington, D.C., 1983 (photo), 101
meeting of American Chemical Society editors, 1978 (photo), 94
Mexico City, 1975 (photo), 80
Mt. Ellen, Henry Mountains, Utah, 1972 (photo), 86
polymer meeting in Moscow, 1960 (photo), 48
Reaction Mechanisms Conference, Northwestern University, 1950 (photo), 42
retirement in Jaffrey, New Hampshire, 1991 (photo), 120
Salt Lake City, 1979 (photo), 100
Samurai warrior in full armor, 1978 (photo), 99
Stanley Steamer (photo), 5
stone lantern factory, Tsukuba, Japan, 1978 (photo), 99
Symposium on Organic Free Radicals, St. Andrews, Scotland, 1984 (photo), 106
Triple Arch, Arches National Park, Utah, 1980 (photo), 85
University of Utah, 1970 (photo), 83
U.S. Rubber Company, 1947 (photo), 37
Utah office staff (photo), 61
Yosemite National Park, California, 1979 (photo), 105
Walling, Cheves (son)
1946 family group (photo), 21
70th birthday symposium, Park City, Utah, 1986 (photo), 107
Mt. Ellen, Henry Mountains, Utah, 1972 (photo), 86
Walling, Frederika C. Haskell (mother), genealogy, 1–2
Walling, Hazel (daughter)
1946 family group (photo), 21
70th birthday symposium, Park City, Utah, 1986 (photo), 107
Walling, Jane (wife)
1946 family group (photo), 21
70th birthday symposium, Park City, Utah, 1986 (photo), 107
Egyptian nightlife, 1974 (photo), 98
Fairbanks, Alaska, 1967 (photo), 70
family life, 21–22
Lesser Sunda Islands, Indonesia, 1989 (photo), 110
Mexico City, 1975 (photo), 80
Park City, Utah, 1980 (photo), 59
retirement in Jaffrey, New Hampshire, 1991 (photo), 120
Samurai warrior in full armor, 1978 (photo), 99
Symposium on Organic Free Radicals, St. Andrews, Scotland, 1984 (photo), 106
Walling, Rosalind (daughter)
1946 family group (photo), 21
70th birthday symposium, Park City, Utah, 1986 (photo), 107
Walling, William English (uncle), National Association for the Advancement of Colored People, 3
Walling, Willoughby G. (father)
American Red Cross, 3
banker, 3
genealogy, 1–2
Walling children, changing a tire in Nevada desert, 1958 (photo), 77
Walton, Sadie, Mexico City, 1975 (photo), 80
Waters, Mrs., Oxford University, England, 1960 (photo), 78
Waters, W. A.
free radical chemistry in Britain, 38
Oxford University, England, 1960 (photo), 78
Webster, Roderick, Stanley Steamer (photo), 5

Westheimer, Frank
luncheon of physical organic chemists, Washington, D.C., 1983 (photo), 101
molecular mechanics, 16
Reaction Mechanisms Conference, Colby Jr. College (photo), 17
Wheland, George, electron distribution and resonance, 16
Wiig, Ed, Fairbanks, Alaska, 1967 (photo), 70
Wiig, Mrs., Fairbanks, Alaska, 1967 (photo), 70
Will, Fritz, director of NCFI, 111–112
Wilson, C., 3rd Reaction Mechanisms Conference, Northwestern University, 1950 (photo), 42
Winstein, Saul, 3rd Reaction Mechanism Conference, Northwestern University, 1950 (photo), 42

Y

Young, Hobart P., Jr.
basement chemistry lab, 5–6
career, 10
Stanley Steamer (photo), 5

Z

Zavitsas, Andreas, Australian National University cheery greeting, 1989 (photo), 103
Zhao, Chengxue
cleavage of complex side chains, 92
Great Wall of China, 1982 (photo), 101
SET reaction, 62
travel in China, 102
Utah office staff (photo), 61

Copy editing: A. Maureen Rouhi
Production: Catherine Buzzell
Indexing: Colleen P. Stamm

Production Manager: Cheryl Wurzbacher

Printed and bound by Maple Press, York, PA